Criminal Justice
Recent Scholarship

Edited by
Marilyn McShane and Frank P. Williams III

A Series from LFB Scholarly

Adolescent Online Victimization
A Test of Routine Activities Theory

Catherine Davis Marcum

LFB Scholarly Publishing LLC
El Paso 2009

Library of Congress Cataloging-in-Publication Data

Marcum, Catherine Davis, 1980-
  Adolescent online victimization : a test of routine activities theory /
Catherine Davis Marcum.
     p. cm. -- (Criminal justice : recent scholarship)
  Includes bibliographical references and index.
  ISBN 978-1-59332-345-5 (alk. paper)
  1. Internet and teenagers. 2. Internet--Safety measures. 3.
Internet--Social aspects. 4. Teenagers--Crimes against. 5.
Technology--Social aspects. I. Title.
  HQ784.I58M36 2009
  364.16'80835--dc22

                                        2009018624

ISBN 978-1-59332-345-5

Printed on acid-free 250-year-life paper.

Manufactured in the United States of America.

Table of Contents

List of Tables

Acknowledgments

Throughout the course of this project, I was extremely blessed to have the suggestions, contributions, and guidance of great minds in the field. First and most importantly, thank you to Dr. David Myers. You have been of undescribable assistance to me and I appreciate the endless time you spent working with me on this project. I will always hold you in the highest regard and appreciate you. Second, a large thank you to Dr. Dennis Giever, Dr. Daniel Lee, and Dr. Jennifer Gossett, for also spending many hours reading this and giving me excellent suggestions for improvement. I appreciate all of your hard work and concern. Finally, thank you to George Higgins for your never-ending guidance and support for the past decade. You have been a wonderful friend and colleague.

I was also extremely blessed to not only have professional support, but personal support as well. Jeff, you have always been a constant source of love, inspiration, and encouragement. Thank you for everything and I love you. Also, a big thank you to my family, friends, and colleagues for their important contributions and support.

CHAPTER 1:
Introduction

In the early 1960s, the idea of an electronic global communication system originated from J.C.R. Licklider of the Massachusetts Institute of Technology (Licklider & Clark, 1962, as cited in Leiner et al., 2003). His "Galactic Network" idea entailed an internationally connected set of computers that allowed for easy accessibility to information. Now known as the Internet, this intercontinental information highway has enabled people of all ages, especially youth, to drastically expand their social circles and improve their ability to communicate with friends and family (Roberts, Foehr, Rideout, & Brodie, 1999; Rosenbaum, Altman, Brodie, Flournoy, Blendon, & Benson, 2000). Approximately half of the teenagers surveyed as part of the Pew Internet and American Life Project stated that the Internet has improved their relationships with friends and family, through almost effortless online communication without the hassle of slower methods, such as the postal service or the telephone (Lenhart, Rainie, & Lewis, 2001). Besides communication and socialization activities, youth are able to participate easily in other activities, such as research, shopping, and online gaming.

Unfortunately, young Internet users are often unable to participate in online activities without the annoyance of uninvited communication from other online users. Furthermore, the activities of youth on the Internet often are disrupted by unwanted sexual and non-sexual harassment by Internet predators, and the frequencies of these types of victimization are increasing (Mitchell, Finkelhor, & Wolak, 2003; O'Connell, Barrow, & Sange, 2002; Sanger, Long, Ritzman, Stofer, & Davis, 2004; Wolak, Finkelhor, & Mitchell, 2004; Wolak, Mitchell, & Finkelhor, 2002; Wolak, Mitchell, & Finkelhor, 2006). A brief summary of the past empirical research performed on adolescent

Internet victimization appears below, to demonstrate the increasing problems youth are encountering while online.

PAST EMPIRICAL RESEARCH

Past empirical research on adolescent Internet use has demonstrated that Internet use by youth has increased drastically in the past 10 years (Addison, 2001; Izenberg & Lieberman, 1998; Lenhart, Rainie, & Lewis, 2001; Nie & Ebring, 2000; Rainie, 2006; United States Department of Commerce, 2002). Rainie (2006) reported recently that 87% of youth under the age of 18 are using the Internet, which is the highest reported statistic to date regarding the prevalence of online use. In the year 2000, 45 million children had Internet access, representing a 700% increase from 1997. These children, ages 2 to 17, experienced the largest increase in the at-home Internet population as compared to any other age group (Addison, 2001).

The various mediums of communication available on the Internet have been a contributing factor to increased Internet use, by providing effortless means of socializing for young people (Clemmitt, 2006; Graham, 2003; Kirkpatrick, 2006; Lamb & Johnson, 2006; Lea & Spears, 1995; Marriott, 1998; Ramirez, Dimmick, & Lin, 2004; Rosen, 2006; Simon, 2006; Stuzman, 2006; Turkle, 1995). The mediums of communication available on the Internet, often referred to collectively as social technology (Lamb & Johnson, 2006), have enabled people of all ages (especially youth) to expand their social circles and improve their ability to communicate with friends and family in an inexpensive manner (Roberts, Foeher, Rideout, & Brodie, 1999). Social technology generally refers to computer-mediated communication (CMC) devices that connect people for personal and professional information sharing. The use of CMC methods allows for ease in the workplace, educational setting, or home to communicate effortlessly with others (Simon, 2006). Although there are numerous ways to communicate and socialize with CMCs, this study will focus on the following mediums: chat rooms, instant messaging, email, and social networking websites. Unfortunately, along with the beneficial use of these CMC methods comes the increased possibility of online victimization.

Several past studies of Internet use by youth have revealed that increasing numbers of young people are experiencing the following types of victimization while using CMC methods: unwanted exposure to sexual material, sexual solicitation, and unwanted non-sexual harassment (Mitchell et al., 2003; Mitchell, Finkelhor, & Wolak, 2007; O'Connell et al., 2002; Quayle & Taylor, 2003; Sanger et al., 2004; Wolak et al., 2002; Wolak et al., 2003; Wolak et al., 2004; Wolak et al., 2006; Wolak, Mitchell, & Finkelhor, 2007; Ybarra, Mitchell, Finkelhor, & Wolak, 2007). The majority of these studies are descriptive in nature, and there is a lack of rigorous research that indicates what online behaviors may increase the likelihood of victimization. Of the few explanatory studies performed, use of chat rooms, discussion of sexual topics with online contacts, and a tumultuous relationship with family or friends have been noted to increase the odds of online victimization (Mitchell et al., 2007; Wolak et al., 2007; Ybarra et al., 2007).

While youth are becoming more susceptible to victimization while using the Internet, few studies have been initiated to examine the adult predators preying on these young people (Armagh, 1996; Beebe et al., 1998; Danet, 1998; Dean, 2006; Durkin, 1997; Durkin & Bryant, 1999; Henderson, 2005; Jones, 1999; Lamb, 1998; McFarlane, Bull, & Rietmeijer, 2000; Quayle & Taylor, 2003; Wolak et al., 2002). Although parents and children are aware of online predators, whether from media reports or actual experience with victimization, they also appear to be less than fully aware of protection mechanisms and efforts, perhaps due to a lack of available information. While government legislation and private entities have made attempts to protect youth from these predators, the problem continues to exist in cyberspace (Fitzpatrick, 2006; Gallo, 1998; Henderson, 2005; Hunter, 2000; Kendall, 1998; McCabe, 2000; Medaris & Girouard, 2002; Mitchell et al., 2005; Mota, 2002; Volokh, 1997).

As stated before, there are few explanatory studies in the literature that attempt to assess causal factors of online victimization and formation of relationships with online contacts. Currently, there are no published studies that use a strong theoretical basis to examine these online outcomes. In the next section, a brief summary will be provided of the theoretical framework used in the present research to better investigate contributory factors that increase or decrease the likelihood of online victimization and formation of relationships with online contacts.

THEORETICAL FRAMEWORK

From the available literature, it is obvious that youth increasingly are becoming victims of online harassment in various forms; however, there is a lack of explanatory studies that assess the factors impacting victimization. Roncek and Maier (1991) suggested that Routine Activities Theory is excellent for use in the examination of predatory or exploitative crimes, which is precisely the type of deviant behavior examined in this study. According to Routine Activities Theory, three elements must be present in order for a crime to occur: exposure to motivated offenders, a suitable target, and a lack of capable guardianship (Cohen & Felson, 1979). This particular study will examine youth Internet use through variables representing the three constructs of Routine Activities Theory.

Routine Activities Theory has proven itself to be useful in explaining different types of criminal victimization. It has been tested and supported by several studies that have examined victimization on the macro-level (Cao & Maume, 1993; Cook, 1987; LaGrange, 1999; Roncek & Bell, 1981; Roncek & Maier, 1991; Sampson, 1987; Tseloni, Wittebrod, Farrell, & Pease, 2004). A more voluminous amount of supporting literature is based on micro-level research, which has examined individual offending behaviors (Bernburg & Thorlindsson, 2001; Felson, 1986; Horney, Osgood, & Marshall, 1995; Sasse, 2005; Schreck & Fisher, 2004), personal and property crime victimization (Arnold, Keane, & Baron, 2005; Cohen & Cantor, 1980; Cohen & Felson, 1981; Collins, Cox, & Langan, 1987; Gaetz, 2004; LaGrange, 1994; Lasley, 1989; Lynch, 1987; Moriarty & Williams, 1996; Mustaine & Tewksbury, 1999; Schreck & Fisher, 2004; Spano & Nagy, 2005; Tewksbury & Mustaine, 2000; Wooldredge, Cullen, & Latessa, 1992), domain-specific models (Ehrhardt-Mustaine & Tewksbury, 1997; Garofalo et al., 1987; Lynch, 1987; Madriz, 1996; Wang, 2002; Wooldredge et al., 1992), and feminist interpretations (Mustaine & Tewksbury, 2002; Schwartz & Pitts, 1995; Schwartz, DeKeseredy, Tait, & Alvi, 2001).

Early studies of Routine Activities Theory focused on the importance of the environment as a necessary component between criminal offenders and victims. This is especially relevant to this study, as the environment of cyberspace is a required component of online victimization and offending. Cyberspace, which thrives on the possibilities of the unknown, provides the opportunity for engaging in activities without the presence of a capable guardian. This applies to both the offender and victim, as both parties potentially can participate in or experience deviant behaviors without much guardianship being present (Beebe et al., 1998; Danet, 1998; Jones, 1999). In general, a lack of behavioral controls encourages willingness to participate in criminal activity, and motivated offenders will place themselves in areas that have an abundance of suitable targets. Further empirical testing of Routine Activities Theory was obtained through this study, by examining the presence of the above-mentioned components of the theory in the realm of cyberspace.

PURPOSE OF THE STUDY

The purpose of this study was to investigate Internet usage in a sample of college freshmen, as well as to consider their experiences with online victimization and the formation of relationships with online contacts. In order to more fully examine this topic area, the chosen methodology was developed under the concepts and propositions of Routine Activities Theory, which has been utilized many times in the past to explain various types of victimization. This study employed a survey and was anticipated to produce a more complete understanding of adolescent Internet use and victimization.

The survey questioned respondents on past and present Internet use and other relevant constructs. Specifically, surveys were administered to enrolled freshmen in the spring of 2008, with a focus on their frequency and types of Internet use, experiences with different types of Internet victimization, and the development of relationships with online contacts. Respondents were asked about their experiences and behaviors when they were high school seniors, as well as their current experiences and behaviors as college freshmen. It was generally hypothesized that spending higher amounts of time on the Internet in certain locations and providing personal information to online contacts increases the likelihood of victimization and formation of relationships with online contacts, while various protective measures decrease the likelihood of these outcomes.

Since the survey to be used was an original instrument that never had been tested, a pilot study was performed to properly evaluate the survey. The survey initially was distributed to four sections of an undergraduate research methods class offered to criminology undergraduates in the fall 2007 semester. Eighty-three completed surveys were obtained and evaluated for reliability and validity purposes. After the survey was amended to improve a few deficiencies, it was distributed to randomly selected freshmen classes, and a sample of 483 surveys was obtained.

Data obtained through administration of the survey were analyzed through various techniques. Frequencies and descriptive statistics initially were produced for all variables to provide an informational summary of characteristics of the sample. Correlation matrices then were prepared for the independent variables representing each theoretical construct, as well as the dependent variables, to consider statistical associations between the variables. Since the dependent variables subjected to further analysis were measured as a dichotomy, logistic regression models were employed to assess relationships between the independent variables and the likelihood of victimization and formation of relationships with online contacts. In addition to the models estimated for the entire sample, split models were used to compare males and females.

It was expected that this study would produce a significant contribution to the literature on adolescent online victimization and relationship formation, considering the overall lack of research on this topic. Explanatory studies on the topic have been rare and limited, and there is a lack of literature using any type of theoretical framework in examining causal factors associated with adolescent online victimization and formation of relationships with online contacts. The main purpose of this study was to identify the existence of possible causal factors, and, in turn, guide policies and programs to educate and protect adolescents online.

The next chapter provides an examination of the current Internet behaviors and experiences of adolescents. This includes a brief overview of the development of the Internet, leading to the current mediums of communication available for use online, as well as a review of activities performed by adolescents and their experiences with unwanted sexual harassment and other victimization while using the Internet. A summary of relevant legislation and preventative efforts by the federal government also will be presented. Chapter 3 presents a description of Routine Activities Theory, along with a review of the existing empirical literature (including macro- and micro-level studies) that supports the use of the theory in examining predatory or exploitative crimes. In Chapter 4, the specific methods used in the study will be explained in detail. Chapter 5 provides the univariate and bivariate results of the study, and Chapter 6 presents the multivariate results. Finally, Chapter 7 contains the conclusions of the study, including a summary and discussion of the findings, policy implications, methodological limitations, and suggestions for future research.

CHAPTER 2:

Adolescents and the Age of the Internet

William Gibson (1984) predicted in his novel, *Neuromancer,* that society's increasing fascination and dependence on computer technology would create a completely electronic world he termed "cyberspace." Cyberspace would be composed of millions of different outlets of information that were easily accessible at the click of a button. However, Gibson also accurately predicted that his new concept would contain dangerous channels leading to sources of vulgarity, criminal activity, and a dangerous hidden world of exploitation. Elmer-DeWitt and Bloch (1995) later insisted that the Internet is full of sexual material, and it is impossible to avoid its exposure. Although awareness is growing, Medaris and Girouard (2002) more recently asserted that youth and parents still are not fully informed of the dangers online and the possible consequences of providing personal information to Internet predators.

This chapter first provides a brief description of the origin of the Internet and several of the current mediums of communication available for online users. Second, since this study examines the Internet habits of young people, past studies investigating Internet use by this target group will be discussed. Another component of the current research is to investigate the occurrence of online harassment by Internet predators; hence, the limited amount of literature on the characteristics of adult predators and their Internet habits will be presented. Next, the association between these two groups will be reviewed with a thorough examination of studies surveying adolescents and their exposure to unwanted sexual harassment and victimization. Finally, a glance at recent legislation and preventative efforts by the federal government will be presented. The goal of this review is to provide a background on the development and change in these arenas and establish the value of the current study.

ORIGIN OF THE INTERNET AND ITS MEDIUMS OF COMMUNICATION

The first recorded notes birthing the idea of the global information system now known as the Internet were written in 1962 by J.C.R. Licklider of the Massachusetts Institute of Technology (Licklider & Clark, 1962, as cited in Leiner et al., 2003). His "Galactic Network" idea entailed an internationally connected set of computers that allowed for easy accessibility to information. While working at the Defense Advanced Research Projects Agency (DARPA), he encouraged the importance of his networking idea to co-workers Ivan Sutherland, Bob Taylor, and Lawrence Roberts. In 1964, fellow MIT colleague Leonard Kleinrock persuaded Roberts that using packets in networking, rather than the commonly used circuits, would produce a more efficient connection. Circuits were only useful for synchronized one-to-one communications. Packets could be used to allow many computers to exchange information asynchronously as peers, independent of each other. The next year, Roberts tested this idea by connecting a TX-2 computer in Massachusetts with a Q-32 machine in California, by using a low speed dial-up telephone line. The confirmation that computers could work together to retrieve information and run programs was sealed, but a circuit telephone system was inadequate for the job of sharing information between computers. However, Kleinrock's declaration of the need for packet switching was reinforced, and the creation of the Internet was on its way (Kleinrock, 1961; Kleinrock, 1964).

Shortly thereafter, Roberts (1967) published his idea for "ARPANET" while working with DARPA. In December 1968, Frank Heart, of Bolt, Beranek, and Newman, began work on the architectural design of ARPANET. One of its major components was the packet switches called Interface Message Processors (IMPs). IMPs, the ancestor of Internet routers, were used by computers to connect to ARPANET with a high-speed serial interface. While the team from Bolt, Beranek, and Newman worked with IMPs, Lawrence Roberts and his team focused on network topology and economics (Leiner et al.,

2003; Roberts, 1967). Leonard Kleinrock's Network Measurement Center at UCLA then was selected to be the first test site for the IMP and host computer connection in September 1969. By the end of the year, four independent host computers were successfully connected into ARPANET, and many more were quickly added over the following years (Leiner et al., 2003).

ARPANET quickly evolved into what is now known as the Internet. Kleinrock's packet switching network was transformed into packet satellite networks; a new version of protocol called Transmission Control Protocol/Internet Protocol (TCP/IP) was developed; and the general public increased its growing demand for computers. By the 1980s, more vendors were incorporating TCP/IP into their products, because of the increased use of networking by businesses and service providers. This in turn heightened interest among private Internet users (Leiner et al., 2003).

With this amplified popularity for technology, the Internet experienced the perfect environment to thrive, and soon began to do so. The goal of the Internet, originating with ARPANET, was to become a collection of communities that provided useful information to its users. By the early 1990s, use of the Internet became a familiar facet in businesses and homes, and by the year 2001, over half of the United States population included regular users of the Internet (Sanger, Long, Ritzman, Stofter, & Davis, 2004). Today's Internet now allows people to shop, make travel arrangements, buy stocks, and most importantly, communicate.

Online Mediums of Communication

The medium of communication on the Internet, often referred to collectively as social technology (Lamb & Johnson, 2006), has enabled people of all ages (especially youth) to expand their social circles and improve their ability to communicate with friends and family in an inexpensive manner (Roberts, Foeher, Rideout, & Brodie, 1999). Social technology generally refers to computer-mediated communication (CMC) devices that connect people for personal and professional information sharing. The use of CMC methods allows for ease in the workplace, educational setting, or home to communicate effortlessly with others (Simon, 2006). Although there are numerous ways to communicate and socialize with CMCs, this study will focus on the following mediums used most often by the population examined in this study: chat rooms, instant messaging, email, and social networking websites.

Chat rooms

Chat rooms allow for individuals to communicate with persons across the county, state, or even the world. These rooms, which usually allow for approximately 25 people to participate in conversation, allow for connections between friends or strangers. Some chat rooms are formed for persons with similar interests, such as Star Trek fans or a sport, while others engage in general socialization. According to Marriott (1998), 40 to 50 million individuals utilize chat rooms. One appeal of chat rooms is that a participant can assume any identity he or she chooses, based on such characteristics as gender, age, and occupation. Online, a person's true identity is hidden and can be amended to the desire of the user (Beebe, Asche, Harrison, & Quinlan, 2004; Freeman-Longo, 2000; Wolak, Mitchell, & Finkelhor, 2002). However, the problem with this option of varied identity is that fellow chat room participants also are afforded that luxury and therefore cannot be trusted in their claims of identity (Lea & Spears, 1995; Turkle, 1995).

Chat room use has been assessed through the Minnesota Student Survey (MSS), which has been administered to Minnesota public school students in grades 6, 9, and 12 triennially since 1989 (Beebe et al., 2004). A particular analysis of the data from 2001 analyzed the 9th grade portion of the sample (n = 50,168). Four types of variables analyzed in the study were: (1) demographics, (2) psychological and environmental factors, (3) behavioral risk factors, and (4) chat room and other Internet activities. The purpose of this research was to identify aspects of Internet use at home associated with the 9th graders in the sample, examined separately by gender, through frequency distributions and the Breslow-Day test of homogenous odds ratios (Beebe et al., 2004).

Of the sample, slightly more than half were girls (50.9%) and the majority was white (82%). Approximately 80% of the 9th graders reported use of the Internet at home, and 48.3% of that group also accessed chat rooms. The results of the study further indicated that chat room use was consistently and significantly associated with risky and antisocial behaviors (i.e., substance use, sexual activity, and truancy), as well as various psychological and environmental factors (i.e., low self-esteem, loneliness, and lack of supervision by parents). Beebe et al. (2004) suggested that this pattern might reflect chat rooms as an alluring experience based on the appeal of the unknown.

Instant messaging

Launched in late 1996, instant messaging (IM) revolutionized the process of online communication (Tyson, 2004, as cited in Flanigan, 2005). It is the process of information exchange through the typing of messages into a text box, which can be seen on the bottom of the text box screen, and then electronically sending it to another user. The response of the other online participant can be viewed on the top portion of the text box screen, which is also where the user can view the entire conversation transcript. Multiple messages can be written and sent simultaneously by both parties (Simon, 2006). This mode of communication is extremely useful, as it allows the immediate transfer of information for business or personal needs (Ramirez, Dimmick, & Lin, 2004). The use of this CMC has become very popular, and the top four IM providers have claimed use by 270 million people worldwide (Graham, 2003). IM also facilitates the exchange of approximately 582 billion messages each day (Radicati Group, 2003). The United States of America alone has approximately 60 million users (Whelan, 2001), 30% of which are between the ages of 18 and 29 (Radicati Group, 2003; Whelan, 2001). According to Olivia (2003), IM use has increased faster than the use of email (33% between 2000 and 2002).

According to Pew Research (2003), IM is utilized almost twice as much by youth than adults (74% versus 44%, respectively). Approximately 13 million adolescents (ages 12 to 17 years old) use IM, 20% of which claim it as their main tool of socializing with friends (Pew Research, 2001). A study of undergraduate college students at a northeastern university (N = 271), designed to assess the use of Internet technologies and their ability to satisfy the needs of the students, found that 92% of the respondents used IM (Ramirez et al., 2004). Participant ages ranged from 18 to 27, with 85% of the sample included in the 18 to 20 year old age range. The results demonstrated that IM was used more than email to meet student needs, and that subjects believed IM was more effective than email for socialization and utilitarian uses (Ramirez, et al., 2004; Flanagin, 2005).

Electronic mail

Electronic mail, or email, has the same premise as traditional mail that goes through the United States Postal Service. However, this method of communication does not require postage and is sent to one or more recipients electronically through the Internet. This transfer of information can take between a few seconds and a few minutes, depending on the type of Internet service and connection. The date of the message and the address of the sender are provided to the recipient upon delivery of the message (Turi, 1997).

Email messages do not require proper letterhead or format, as do paper-based business communications. Also, the language used in emails, such as abbreviations to represent phrases and symbols to represent emotions, can influence electronic communication to be more informal and less restricted; therefore, communication in email can be radically different than face-to-face communication. McCarty (1990, as cited in Turi, 1997) argued that the removal of social constraints and the ability to be freer in expression could change personae in the communicators. For example, McCarty refers to "flaming," which is an extreme outburst of emotion often present in emails. A face-to-face communication would be more likely to produce reserved conversation, because of the expected social cues in these conditions.

Social networking websites

Interactive social network websites, such as MySpace and Facebook, allow members to utilize internal email, blogs, personal profiles, videos, music, and photographs to communicate with others. Personal profiles contain a wealth of information, including personal calendars, interests, and location. Many of these websites, such as MySpace.com, have open membership; therefore, profiles can be viewed by anyone online. Facebook, which has approximately 8 million members, requires formal membership before access is granted to review profiles (Clemmitt, 2006).

Currently, the largest of these networking sites is MySpace.com, which has approximately 80 million members (PRNewswire.com, 2006). According to Rosen (2006), users typically access MySpace.com for two hours a day, five to seven days a week. Between April 2005 and April 2006, use of the site jumped by 367%, and overall use of the top 10 social network websites increased by 47% (Kirkpatrick, 2006).

MySpace.com recently has come under harsh media criticism for its blatant exposure of personal information pertaining to a vulnerable population: children. MySpace.com allows users as young as 14 to register a profile, but users under the age of 16 automatically experience restrictions on the availability of information to the general public (MySpace.com, 2006). Also, when registering a profile, the "Terms & Conditions Agreement" warns of providing personal information to other users. However, it is quite easy to consent to this agreement and disregard the warning, as well as register under a false age to obtain full privileges and rights to information exposure. To illustrate, in early 2006, a 14 year-old Texas girl and her mother filed a lawsuit against MySpace.com, claiming the website should claim responsibility for the sexual assault of the girl by a man she met on the website (Osborn, 2006). The lawsuit claims that MySpace.com should share blame based on the fact that the security measures to prevent contacts with children under the age of 16 are useless, because users are not required to verify their age.

Characteristics of Adolescent Internet Use

Since its inception and growth into mainstream practice, the use of the Internet has transformed from purely governmental operations to a plethora of functions for personal use by adults and children. Rainie (2006) reported that currently 87% of youth under age 18 are using the Internet, the highest reported statistic to date regarding frequency of online use. These young people are performing a variety of functions online, including research, shopping, and communication.

The age of first use of Internet users also is decreasing, while the actual amount of time spent online is steadily increasing (Addison, 2001; Izenberg & Lieberman, 1998; Nie & Erbring, 2000; United States Department of Commerce, 2002). From 1998 to 2001, the use of the Internet by three and four year olds grew from 4.1% to 14.3%. In all other age groups (5-9 years, 10-13 years and 14-17 years), usage jumped by at least 20% between the years of 1998 to 2001 (United States Department of Commerce, 2002). In the year 2000, 45 million children had Internet access, representing a 700% increase from 1997. These children, ages 2 to 17, experienced the largest increase in the at-home Internet population compared to any other age group (Addison, 2001).

The use of the Internet on college and university campuses has increased quickly as well. Student culture now revolves around the use of the Internet. College classes are being taught completely online with the use of discussion boards and chat rooms. Social activity, sporting events, and academic announcements are made over email. Students and professors often provide the preferred method of contact as an email address or personal web page, as opposed to telephone. Kendell (1998) attributes this dependency on technology as being a result of the expectation in academia to stay on the forefront of technological advances; consequently, students are expected to maintain an advancing level of knowledge.

Numerous studies have been conducted to examine the frequency and purposes of adolescent Internet use (Beebe et al., 2004; Lenhart, Rainie, & Lewis, 2001; Mitchell, Finkelhor, & Wolak, 2003; United States Department of Commerce, 2002). In 2000, the Pew Internet & American Life Project performed a tracking poll of households in the United States with Internet use. Of those households that reported having children, a sample was randomly selected to complete a further telephone survey on the frequency of adolescent Internet use (Lenhart et al., 2001). Some of the households selected for the survey were unable to be reached or chose not to participate (the percentage of non-participants was not noted in the study); therefore, this survey cannot be considered as containing a representative sample of the United States adult and adolescent population.

The actual survey sample included 754 adolescent Internet users, as well as one of the parents or guardians who lived with them in their household. According to Lenhart et al. (2001), the adolescent subjects reported the majority of their use time was spent sending or reading email (92%), sending Instant messages (74%), and visiting chat rooms (55%). A second major purpose for their reported use was conducting research. Sixty-eight percent of the adolescent sample relied on the Internet for their source of daily news, while 66% used it as a source for information on a certain product or service. Finally, the youths in the sample also noted their use of the Internet for entertainment purposes. A large percentage (83%) reported visiting entertainment sites as a regular activity, as well as listening to or downloading music (59%).

Recent research also has indicated that youths who experience loneliness and isolation in their lives often form online relationships with strangers (Lorig et al., 2002; Reeves, 2000; Wang & Ross, 2002; Wolak, Mitchell & Finkelhor, 2003). The first Youth Internet Safety Survey, sponsored by the National Center for Missing and Exploited Children, is a nationally representative study of 1501 youth (796 boys and 705 girls) who were 10 to 17 years old and participated in regular use of the Internet (regular Internet use was defined as "using the Internet at least once a month for the past six months on a computer at home or some other location") (Wolak et al., 2002, 442). This study determined that over half of the youth (55%) examined reported the use of chat rooms, instant messages, and email to communicate with people they had never met, with the hopes of forming relationships (Wolak et al., 2002). Various relationship types were reported as forming through these communications, including casual online friendships (25%), close online friendships (14%), face-to-face meetings (7%), and online romantic relationships (2%). Females were more likely than males to form close online relationships (19% versus 16%, respectively), which included a close friendship, romance, or face-to-face meeting. Only two youth reported discomfort after face-to-face meetings with their online friend, and only one of these meetings involved a youth and an adult. Specifically, when a 16-year-old girl met her online male friend in his thirties, she became uncomfortable after he invited her back to his hotel room. She declined his request and immediately left the scene (Wolak et al., 2002).

This study also revealed that level of conflict and communication with parents was a predictor of the likelihood of formation of close online relationships. Bivariate analyses demonstrated that a dis-proportionate number of youth who formed close online relationships had a high amount of conflict with their parents, a low standard of communication, and were involved in delinquent activities. Females who were highly troubled (i.e., high levels of depression and victimization by peers) or reported yelling, nagging, or severe punishment by their parents were twice as likely as the other females in the sample to form close online relationships (Wolak et al., 2003).

Other evidence indicates some younger users of the Internet spend much of their time searching for sexual material. The word "sex" is one of the most frequently searched words on the Internet by adults, as well as teenagers and college students (Cooper, 1997; Freeman-Longo, 2000; Goodson, McCormick, & Evans, 2000; Schnarch, 1997). Adult and child pornography is often a tool of seduction used by online predators (Quayle & Taylor, 2003). Pornographic websites can be accessed easily, even by youth who are underage. Verification of one's age on the Internet is virtually impossible, as no identification typically is required, providing free access to chat rooms, videos, and web sites with adult-oriented materials (Freeman-Longo, 2000).

STUDIES OF ONLINE VICTIMIZATION

Numerous studies in recent years have supported the fact that Internet victimization of youth by online predators is on the rise (Mitchell et al., 2003; O'Connell, Barrow, & Sange, 2002; Sanger, et al., 2004; Wolak et al., 2002; Wolak, Finkelhor, & Mitchell, 2004; Wolak, Mitchell, & Finkelhor, 2006). Young victims often are exposed to explicit sexual material or unwanted sexual solicitation and harassment . This harassment is not limited to a certain gender or age group, as can be seen in the supporting literature, nor does it appear to be decreasing.

Youth Internet Safety Survey

The previously mentioned Youth Internet Safety Survey is a nationally representative study of 1501 youth (796 boys and 705 girls), ages 10 to 17 years old, who participated in regular use of the Internet (Wolak et al., 2002, 442). Sponsored by the National Center for Missing and Exploited Children, the primary purpose of the survey was to assess the frequency of youth experiences with unwanted sexual solicitation, pornography, and harassment online. The first survey administration involved telephone interviews with the youth between August 1999 and February 2000, regarding the amount and types of online harassment they had received over the Internet. Prior to the interview with the youth, a brief interview was conducted with a single parent or guardian of the household to confirm the regularity of Internet use and obtain permission to conduct interviews with the adolescents in the household (Mitchell, Finkelhor, & Wolak, 2003).

The sample employed was not representative of all youth in the United States, because the availability of the Internet was not evenly distributed across the population during the time of the survey. However, the sample was representative of the population of Internet-using youth. Of the sample, 53% were male and 47% were female.

Thirty-eight percent were between the ages of 10 and 13, while the remaining participants were between the ages of 14 and 17. Over 70% of the sample was of non-Hispanic white ethnicity, and almost half lived in a household with an annual income of over $50,000 (Mitchell et al., 2003; Wolak et al., 2002).

The incidences of online victimization by adults were grouped into three categories: 1) sexual solicitation and approaches, 2) unwanted exposure to sexual material, and 3) harassment. One in five of the youths reported receiving an unwanted sexual solicitation in the previous year; 5% of those youth reported a solicitation that actually left them feeling distressed or afraid. Although girls are generally considered to be the main targets of sexual solicitation online, 34% of the males in the sample identified themselves as targets. Over three-fourths of the targeted youth were at least 14 years old. Twenty-five percent of the sample had unwanted exposure to sexual material while surfing the Internet or through opening emails or Instant messages. Males (57%) and youth at least 15 years of age (63%) most often reported unwanted exposure to sexual material. Only 6% of the youth in the sample had received threats or other offensive behavior termed as harassment, with 2% actually feeling some sort of fear or distress as a result of the harassment. The likelihood of harassment for each gender was equal, while 70% of the events occurred with youth 14 years of age or older (Mitchell, Finkelhor, & Wolak, 2003).

During the time period of March to June 2005, a second administration of the Youth Internet Safety Survey (YISS-2) took place. Of the sample of 1500 youth, 49% were males and 51% were females. Thirty-six percent were between the ages of 10 and 13, while the remaining participants were between the ages of 14 and 17. Seventy-six percent of the sample was of non-Hispanic white ethnicity, and more than half (57%) lived in a household with an annual income of over $50,000 (Wolak, Mitchell, & Finkelhor, 2006).

There were several changes regarding the occurrence of harassment in the second study compared to the first. First, the proportion of youth who reported online harassment grew from 6% to 9%. Despite the increased usage of filtering and blocking software by parents (55% of parents in the second study reported the use of this software), unwanted exposure to sexual material increased by 9%. Also, of the unwanted exposure to sexual material, the number of youth who reported distressing reactions to the material grew 3% since the first survey. A larger percentage of youth received unwanted sexual solicitation compared to the first survey (13% in 2001 versus 19% in 2006). However, aggressive solicitations, which included attempts to contact the youth offline, did not increase (Wolak et al., 2006).

A more recent study used data from the YISS-2 to identify online behaviors that increase the likelihood of online victimization. Youth were found to participate in several types of risky behaviors online, such as disclosure of personal information, talking about sex with someone known only online, and harassing others online. Of these risky behaviors, talking about sex with unknown people online and meeting people online in multiple ways were found to produce significantly higher odds of online interpersonal victimization. However, posting or sending personal information online by itself was not significantly associated with increased odds of online interpersonal victimization (Ybarra, Mitchell, Finkelhor, & Wolak, 2007).

As mentioned before, there is a gap in the literature regarding explanatory studies that examine online victimization through the use of advanced statistical analyses. Of the few published studies that use any type of advanced quantitative analysis, data from the Youth Internet Safety Survey were used (Mitchell, Finkelhor, & Wolak, 2007). The first study examined youth Internet users at risk for serious online sexual solicitations. Data from both administrations of the survey were combined to examine changes in the prevalence of youth reporting unwanted sexual solicitation. Two logistic regression models were estimated, including the year of the study as an independent variable and sexual solicitation as the dependent variable. It was determined youth in the second administration of the survey were more likely to report aggressive sexual solicitation (i.e., solicitations which attempted to or established offline contact) than youth in the first administration. However, these same youth were less likely to report online-limited sexual solicitation (Mitchell et al., 2007).

The second portion of the study involved data solely from the second administration of the survey. Incident characteristics between aggressive and online-limited sexual solicitations were compared using chi-square tests. It was determined that aggressive sexual solicitations were more likely to be committed by someone the youth knew, most commonly a male at least 18 years old. If the youth received multiple solicitations, they were more likely to come from the same person. Multinomial logistic regression analysis was conducted to identify factors associated with sexual solicitations. Being female, using chat rooms, and talking with people online (especially about sex) increased the risk for aggressive sexual solicitations more than online-limited incidents (Mitchell et al., 2007).

The second study employing advanced statistical techniques using data from the second administration of the Youth Internet Safety Survey to examine the issue of unwanted and wanted exposure to online pornography, as well as its relationship with online harassment and unwanted sexual solicitations (Wolak, Mitchell, & Finkelhor, 2007). Cross-tabulations were run to determine what characteristics of youth were most associated with exposure to pornography. It was found that the age group of 13 to 17 years reported the most unwanted and wanted exposure to pornography. Bivariate analyses indicated that most of the prevention characteristics (e.g., use of filtering software), psychosocial characteristics (e.g., parent-child conflict), and Internet use characteristics were significantly associated with exposure (Wolak et al., 2007).

Multinomial logistic regression models also were estimated to examine the characteristics associated with exposure to online pornography. Youth who used file-sharing programs or reported harassment or sexual solicitation online had a higher risk of being exposed to unwanted and wanted pornography. Youth who noted indicators for depression also were more likely to report unwanted and wanted exposure to this material. On the other hand, the use of filtering software and monitored Internet use decreased the likelihood of exposure for youth (Wolak et al., 2007).

Further Studies of Online Harassment

The Youth Internet Safety Survey is one of the most noted studies on online harassment, but it is not the only valid empirical study on the subject. Overall, youth are reporting a large amount of harassment during their time online, regardless of the method of CMC used. The research discussed below provides an insightful overview of the experiences facing youth while surfing the Internet.

The Cyberspace Research Unit at the University of Central Lancashire surveyed the use of chat rooms by youth in the United

Kingdom in two different time periods: early 2002 and late 2002. The first sample (n = 1369) was composed of children ages 9 to 16 years old. Twenty percent reported harassment in a chat room by another participant, and 14% actually admitted to harassing someone in a chat room. The study was repeated again later in the year with a sample (n = 1331) of children ages 8 to 11 years old, revealing the same percentages of harassment and harassing behavior. A large portion of this sample (53%) also reported having conversations of a sexual nature online. Both of the groups, despite the differences in ages, had similarities in the frequency of chat room use and their experiences. Nineteen percent of both samples used chat rooms regularly, and 10% of the chat room participants actually engaged in face-to-face meetings (O'Connell et al., 2002).

Another interesting study surveyed an atypical population, as compared to the majority of other online harassment studies, but the results were similar. Sanger et al. (2004) surveyed 62 adolescent females in a juvenile correctional facility regarding their use of the Internet and involvement in chat room conversations. The girls were questioned regarding their choice of chat rooms as a form of communication, approximate number of hours per week in the chat rooms, and potential online harassment. Approximately 87% of the girls (n = 54) reported participation in chat rooms. Seventy percent (n = 38) of the reported participants stated they had been approached for sexual behaviors (Sanger et al., 2004).

Another study of chat room use and abuse involved the observation of America Online's Kids Only and Teen chat rooms for aggressive comments, sexual comments, and cursing. Bremer and Rauch (1998) found that sexual and aggressive comments were more common in the Teen chat rooms (one sexual comment every 4 minutes and one curse every 1.5 minutes) than in the Kids Only chat rooms (one sexual comment every 21 minutes and one curse every 15 minutes). The benefit identified with these chat rooms is that there are monitoring capabilities. Youth in the rooms can block harassing or obnoxious offenders from their screen if they are annoyed or feel uncomfortable with the comments of the offender (Bremer & Rauch, 1998).

Although most studies have involved the investigation of youth under the age of 18, as they are the population that experiences harassment and victimization more than any other age group, college undergraduates also are utilizing the forms of CMCs used by the younger crowd (Harris Interactive, 2001). As a result, they are experiencing forms of online harassment. In April 2002, 339 undergraduates at the University of New Hampshire were surveyed with regard to their experiences with online harassment (Finn, 2004). This included insulting or threatening instant messages or emails, receipt of unwanted pornography, and stealing a person's identity online. Although the study employed a convenience sample, it was collected in a systematic manner. The research group approached every fourth person encountered in the student union on various days and times until an adequate sample was collected. Only 60 (13.7%) of the students approached declined participation.

Students in the sample reported frequent use of the Internet, and 97% reported email use at least one time per week. Eighty-one percent of respondents reported use of an instant message service at least one time per week. With regard to harassment, approximately 60% reported receiving unwanted pornography, and 10% reported receiving threatening instant messages or emails. Only 7% of the students actually reported the harassment to the authorities (Finn, 2004).

Studies of Harassment through Social Networking

The personal information available on individual web pages in social networking sites can be more dangerous that many users perceive. For example, a teenage girl may assume that not posting the name of her school or hometown on her personal profile equates to safety. However, if she has other friends who list that information on their personal page, and she is listed as one of their buddies, she can become a target as well (Collins, 2006, as cited in Clemmitt, 2006).

Stutzman (2006) surveyed freshman at the University of North Carolina regarding the personal information shared on their social networking site page. Almost all reported posting their birthday and hometown on their page (96.2% and 94.2%, respectively). Other facts listed were relationship status (82.3%) and cell phone number (16.4%). This study is relevant to the current research, which also will be examining the types of personal information shared on social networking sites. It will be hypothesized that undergraduates are providing revealing, and potentially dangerous, information about themselves that can be accessed easily by any online user.

Another study of social networking websites was performed by Rosen (2006). Through a convenience sample, he surveyed parents and children in the Los Angeles area regarding their use and knowledge of MySpace.com. Data were collected strictly from children under age 18 in the first sample in March 2006 (n = 1,257), and then again in June 2006 with pairs of teenagers and parents (n = 534). Participants were given a link to a website where they accessed an anonymous survey containing 65 questions regarding demographics, MySpace.com usage and experiences, and psychological variables, including depression, family support, and Internet addiction. With regard to the adolescents, Rosen found that 83% believed MySpace.com was a safe website, and 70% would be comfortable showing their parents their personal page. Thirty-five percent of the youth reported concern regarding sexual predators on MySpace; 48% believed that there were "some" sexual predators on MySpace.com; and approximately 8% reported being approached for a sexual meeting. In contrast, 83% of the parents were concerned about sexual predators on MySpace.com; 63% believed there are "quite a few" sexual predators on MySpace.com; and 81% reported worrying about their child participating in face-to-face meetings with online friends. Ironically, and despite the substantial expression of concern, only 38% of the parents discussed the tenets of MySpace.com with their child (Rosen, 2006).

Involvement of Law Enforcement Agencies

The involvement of law enforcement is vital to increase the level of safety of youth online (Dean, 2006). As online harassment and victimization is becoming a growing problem, the caseload of these crimes experienced by law enforcement agencies also is growing. This growth is demonstrated in recent research.

Between October 2001 and July 2002, a stratified random sample of law enforcement agencies (N = 2574) were surveyed by mail and telephone interview with regard to their experience with sexual offenses against juveniles originating with an online encounter (Wolak, Finkelhor, & Mitchell, 2004). The National Juvenile Online Victimization Study polled federal, state, county, and local law enforcement agencies having a wide range of experience with Internet-related sex crimes against youth. Some agencies had specialized departments exclusively dealing with this type of crime, while others had more limited experience. Of the victims reported by law enforcement agencies (n = 129), 75% were between 13 and 15 years old. Three-fourths of the victims were female. The reported offenders tended to be over the age of 25 (76%) and male (99%), and generally did not deceive their victims regarding the fact they were adults and interested in sexual relationships (Wolak, Finkelhor, & Mitchell, 2004).

This study also revealed the majority of first encounters between offenders and victims occurred in an online chat room (76%). These chat rooms were oriented toward teen interests, dating, and romance, and a few were dedicated to adult and juvenile sexual encounters. Communication extended to telephone conversations for 79% of the victims; 48% received pictures of the offender and 47% received gifts or money. Most of the cases also developed into a face-to-face encounter (74%). Of those meetings, 93% involved illegal sexual conduct between the offender and victim. Some of the sexually abused victims were given drugs or alcohol (40%) or were exposed to adult or child pornography (23% and 15%, respectively). Of those victims not sexually abused, their communications ended with either a report to the police or intervention by a friend or family member (Wolak et al., 2004).

Summary of Adolescent Internet Use and Victimization

Past research has demonstrated that the frequency of Internet use by youth has increased drastically in the past 10 years (Addison, 2001; Nie & Ebring, 2000; Izenberg & Lieberman, 1998; Lenhart et al., 2001; Rainie, 2006; United States Department of Commerce, 2002). Furthermore, the various mediums of communication available on the Internet provide effortless means of socializing for young people (Clemmitt, 2006; Graham, 2003; Kirkpatrick, 2006; Lamb & Johnson, 2006; Marriott, 1998; Lea & Spears, 1995; Ramirez et al., 2004; Rosen, 2006; Simon, 2006; Stuzman, 2006; Turkle, 1995). However, along with the luxury of this advanced technology that enables communication between friends and family, the potential for developing relationships with new online friends, as well as adversaries, is enhanced. Descriptive studies of Internet use by youth also have found that many youth are experiencing victimization online (Mitchell et al., 2003; O'Connell et al., 2002; Quayle & Taylor, 2003; Sanger et al., 2004; Wolak et al., 2002; Wolak et al., 2003; Wolak et al., 2004; Wolak et al., 2006). The few studies available using advanced statistical analysis has indicated that certain characteristics put youth at greater risk for victimization (Mitchell et al., 2007)

From the available literature, it is obvious that youth increasingly are becoming victims of online harassment in various forms. However, what has not been thoroughly explored is the relationship between frequencies and types of Internet usage by youth and their likelihood of victimization. Also, the amount and types of personal information provided by youth to online persons and its relationship to online victimization is in need of further study. Overall, explanatory studies are lacking in the current literature and are needed to better examine the growing occurrence of online victimization. This study will seek to make a contribution to the literature on online victimization by examining these factors and their relationship to the Internet victimization of young adults.

CHARACTERISTICS OF ADULT PREDATORS

Danet (1998) has argued that the anonymity of the Internet could have a dangerous and uninhibiting effect on the behavior of those who use it. People can assume different identities, act in unusual ways, and explore aspects of their own desires and fantasies that would be restricted in their normal lives. Children also can be lured by the excitement of their ability to be a completely new person, as can adults who may be searching for them. Furthermore, the unrestrained ability to participate in sexual deviance that is available to adults with a sexual interest in children may often become too tempting to limit to an online relationship (Beebe et al., 1998; Danet, 1998; Jones, 1999).

Dozens of media reports have portrayed stories of manipulative predators gaining the trust of adolescents online and eventually earning a face-to-face meeting with the intention of a sexual encounter (Armagh, 1996; Wolak et al., 2002). Durkin (1997) has asserted that adults with a sexual interest in children utilize the Internet in four ways to satisfy their sexual desires: trafficking of child pornography; communication with other pedophiles; location of molestation victims; and engagement of inappropriate sexual communication with children. Moreover, as the Internet has developed and provides the ability to participate in a range of activities in perceived privacy, predators now appear bolder when it comes to engagement in illegal sexual behaviors.

According to NetSafe/Internet Safety Group, there are approximately 50,000 sexual predators online at any time (as cited in Dean, 2006). However, few studies actually have examined the adults who commit sexual offenses against children.

Lamb (1998) initiated one of the only systematic studies of chat room use by pedophiles. The purpose of his study was to characterize the personalities and behaviors of chat room visitors who specifically targeted young gay men. Three categories of users were identified: browsers, cruisers, and pornographers. Browsers were genuine people hoping to meet people, but they avoided any type of sexual conversation. Cruisers, the largest group, possessed an extensive knowledge of sexual practices and were eager to discuss their experiences. Many of the cruisers also wished to seek contact outside of the chat room. Finally, pornographers did not participate in personal conversations, but were highly regarded as experts in their field and focused most of their energy on trading photographs. Lamb further concluded from his study that most participants did not restrain themselves from sexual conversations with other chatters they assumed to be children (Lamb, 1998).

The trafficking of child pornography has become a lucrative business that continues to involve a growing number of participants and has gained the attention of law enforcement. In 1998, for example, the federal government disbanded the traffickers of "The Wonderland" group (McAuliffe, 2001). This child pornography group had 200 members in over 40 countries, including the United States, Australia, Germany, Great Britain, and several other European countries. In February 2001, seven Britons were sentenced to between 18 and 30 months for their involvement in the pornography ring. One of the defendants, Antoni Skinner, was found to have approximately 750,000 indecent images on his computer, involving over 1,000 children (McAuliffe, 2001). In addition, one of the most recent major apprehensions involved the arrest of 125 people nationwide for possession and receipt of child pornography. Children as young as 6 months old were featured in photographs and videos showing sexual acts with adults. The men and women arrested included a Bible camp counselor, a pharmaceutical researcher, and a Boy Scout leader (Parry, 2006).

Quayle and Taylor (2003) have argued that many adult predators use child pornography as a form of seduction, and we as a society attempt to understand these adults in a cognitive-behavioral way. In other words, we believe the thinking of the child sexual offender is assumed to be based on beliefs that minimize and rationalize the disturbing behavior. Quayle and Taylor (2003) further asserted that some offenders are unable to connect their mental state as wrong and relate it to the needs and beliefs of other people. However, other offenders are clearly aware of their actions and manipulative practices, as they can describe how a child was targeted and how the child and family were influenced to increase the opportunity for offending (Quayle & Taylor, 2003).

Leigh Baker (2003, as cited in Henderson, 2005), a recognized child abuse expert, constructed a list of 10 general characteristics of sexual offenders. Her list included the inability to form intimate relationships with adults, a history of abuse, deviant sexual behavior and attitudes, and a sense of entitlement. Although these characteristics may be shared by online pedophiles and those sitting on playground benches, their techniques of manipulation are different. Predators that use face-to-face techniques generally gain the affection of children through discussion of common interests. Online predators can assume any identity they choose and easily merge into teen culture without having to display a physical appearance to reveal their adult status (Henderson, 2005).

While some Internet predators choose to actively recruit children for sexual purposes, those who maintain only online communication have been noted to justify their behavior by saying it prevents actual injury to the child (Durkin & Bryant, 1999). Their sexual desires are satisfied from the knowledge that they are discussing sexual acts with a young child, and they do not require further pursuance of any physical act. However, not all predators who start out this way maintain only online communication. In other words, the Internet can facilitate the communication of sexual desires through words, but heated conversations may lead to requests for face-to-face meetings that result in sexual activity (McFarlane, Bull, & Rietmeijer, 2000).

For example, in United States v. Bodenheimer (2005/2006), formerly in the Northern District of Illinois and appealed to the Tenth Circuit Court of Appeals, a 31 year-old man had developed an online relationship with a 13 year-old girl who had been previously molested. After two months, he gained her trust, proposed marriage, and convinced her that he was in love with her. After meeting in Chicago, he brought her to a hotel and engaged in numerous acts of sexual intercourse. Bodenheimer returned home after the meeting and continued communication with the girl, as well as numerous other minors. After the girl's parents reported the incident to the police, Bodenheimer was arrested and charged with traveling in interstate commerce for the purpose of engaging in a sexual act with a person less than 18 years of age (Kendall, 1998).

Similarly, Katherine Tarbox (2000) wrote a book entitled Katie.com, describing her true-life story of meeting a man on the Internet and maintaining contact with him. He gained her trust, assumed a false identity, and proclaimed his feelings for her. Finally, he convinced her to meet him while she was at an out-of-state swim meet. Fortunately, Katie was found by her mother in the room of the man as he was actively attempting to molest and rape her. Her predator was actually 20 years older than he had claimed and had a different name (Tarbox, 2000).

A final note on these adults concerns the nature of their personal characteristics. Mitchell, Wolak and Finkelhor (2005) utilized a sub-sample (n = 124) from the National Juvenile Online Victimization Survey to examine persons arrested for Internet crimes against minors since July 1, 2000. Of the sub-sample, all but one was male, 91% were white, and 91% were employed full time. However, aside from these characteristics, other descriptive qualities (i.e., age, education, and employment type) tended to be very diverse; therefore, online sexual abusers cannot be easily profiled, and they may generally appear to be a respected society member.

LEGISLATION AND PROGRAMS AGAINST SEXUAL EXPLOITATION AND HARASSMENT OF CHILDREN

As can be seen from the literature, online victimization of youth has become a known problem. Federal and state legislators have made vigorous efforts since the 1970s to reduce these crimes through numerous statutes, some of which have been ruled unconstitutional soon after inception. Described below are statutes that have received the most notoriety, as well as other federal attempts to prevent online victimization.

Realizing that child pornography was a profitable venture for pedophiles, in 1977 Congress passed the "Protection of Children against Sexual Exploitation Act" (Simon, 1999). It made illegal the use of a minor younger than age 16 in a visual production of sexually explicit conduct. This age limit was raised to 18 in 1984 (McCabe, 2000). A further amendment contained in the "1986 Child Sexual Abuse and Pornography Act" banned the production and use of advertising for child pornography, and made it a personal injury civil liability for children to be used in child pornography production (Mota, 2002). In 1988, the passing of the Child Protection and Obscenity Enforcement Act made it illegal to transport, distribute, and receive child pornography by computer. Subsequently, the Supreme Court found in *United States v. X-Citement Video* (1994) that the government must prove that the defendant knew the persons pictured in child pornography were minors (Simon, 1999).

In the mid-1990s, Senator Orrin Hatch authored United States Senate Bill 1237, later known as "The Child Pornography Prevention Act (CPPA) of 1996." The bill amended the definition of child pornography to include the photography, filming, and video recording of sexually explicit conduct of real minor children, as well as digitally-created images of a child involved in pornography. The United States Code 18 U.S.C.A. § 2256(8) defined child pornography as:

> any visual depiction, including any photograph, film, video, picture, drawing or computer or computer-generated image or picture, which is produced by electronic, mechanical or other means, of sexually explicit conduct, where: (1) its production involved the use of a minor engaging in sexually explicit conduct; or (2) such visual depiction is, or appears to be, of a minor engaging in sexually explicit conduct; or (3) such visual depiction has been created, adapted or modified to appear that an 'identifiable minor' is engaging in sexually explicit conduct; or (4) it is advertised, distributed, promoted or presented in such a manner as to convey the impression that it is a visual depiction of a minor engaging in sexually explicit conduct (pp. 106).

With this amendment, a person could be charged with possession of computer-generated child pornography (Henderson, 2005; Kendall, 1998; McCabe, 2000).

The constitutionality of the CPPA quickly was challenged. In *United States v. Hilton* (1998), the defendant asked that charges of possession of child pornography violating the CPPA be dismissed. Hilton stated that the statute prohibited constitutionally protected speech by banning adult pornography, and that the language in the statute was vague and overbroad. Although the United States Supreme Court found his first claim to be unmeritorious, it did rule that the language of the CPPA was vague and did not clarify the prohibited conduct (Simon, 1999).

The constitutionality of the CPPA was again challenged by a group of plaintiffs known as the Free Speech Coalition. The United States District Court of the Northern District of California, in *The Free Speech Coalition v. Reno* (1997), ruled that because the CPPA does not require advanced approval for the production of adult pornography that does not include minors, nor does it entail a complete ban on constitutionally protected material, it is not a violation of the First Amendment. However, in 1999 the Court of Appeals for the Ninth Circuit reversed the district court's ruling on the Act. It ruled that the First Amendment does not allow Congress to pass a statute that criminalizes a mere generation of an image, because an actual human being was not involved (Mota, 2002). In 2002, the United States Supreme Court held that since actual children were not used in the photographs and videos in question, these productions were protected by the First Amendment (Henderson, 2005).

The "Communications Decency Act" (CDA), a part of the "Telecommunications Act of 1996," was enacted to limit the exposure of children to sexually explicit pictures available online (Mota, 2002). Under this Act, any person who knowingly creates, solicits or transmits images to a minor under the age of 18 could be penalized by imprisonment for up to two years and/or a fine of $250,000 per offense. However, on June 26, 1997, the United States Supreme Court decided that the CDA's "indecent transmission" specification violated the First Amendment's guarantee of free speech. Telephone companies and Internet providers were declared not liable for indecency that was beyond their control. In response to the dismissal of the CDA,

Congress then passed the "Child Online Protection Act" (COPA) (Henderson, 2005; Hunter, 2000; Mota, 2002).

COPA used contemporary standards based on a test developed in *Miller v. California* (1973). The *Miller* test contains three components that are used to determine if speech is obscene: 1) whether the average person would contend that the material contains prurient value; 2) whether the work depicts sexual acts or excretory functions in an offensive way; and 3) whether the material lacks serious artistic, literary or political value (Virginia Tech, 1997). COPA applied only to material put on the Internet and made for commercial purposes, and it restricted only the documentation harmful to minors. Under this Act, any accused persons must have knowledge of the content of the material, and the material must meet the standard of appealing to the prurient interest of an average person. A violation was classified as a misdemeanor, with the punishment of six months in jail and a $50,000 fine for each violation. The Attorney General also was authorized to collect $50,000 in civil penalties. However, in 2004, the United States Supreme Court ruled in *Ashcroft v. American Civil Liberties Union* (2002) that once again, the tenets of the Act were in violation of the First Amendment (Henderson, 2005; Hunter, 2000; Mota, 2002).

The Children's Internet Protection Act (CIPA) was enacted by Congress in December 2000 and addressed access to offensive Internet material on school and library computers. Certain requirements of CIPA were imposed on schools and libraries receiving federal funding for Internet access under the E-rate program; if these regulations were not followed, the E-rate funding could be removed. The CIPA regulations basically required filtering and blocking software, as well as other safety measures, to protect children from accessing obscene and harmful material (Federal Communications Commission, 2006). Building on the concept of CIPA, the Deleting Online Predators Act (DOPA) was introduced to the House of Representatives in spring 2005. DOPA intensifies the regulations of CIPA, as it would require schools and libraries to completely restrict children's access to all Internet sites through which strangers can contact them (Fitzpatrick, 2006). Currently, it has passed the House of Representatives by roll call vote and is awaiting Senate approval.

One of the newest pieces of federal legislation is the Adam Walsh Child Protection and Safety Act of 2006. According to the Bush administration ("Fact Sheet: The Adam Walsh Protection and Safety Act of 2006," 2006), this Act will strengthen and supplement previous government actions to protect children online in four ways: 1) expanding the National Sex Offender Registry; 2) strengthening federal penalties for crimes against children; 3) making it difficult for sex predators to have access to children on the Internet; and 4) creating a National Child Abuse Registry and requiring in-depth background checks of adoptive and foster parents before granting custody of a child. The main goal of the Act is to strengthen current federal laws in terms of harsher punishments for online predators, as well as provide law enforcement with more tools to investigate and apprehend these offenders ("Fact Sheet: The Adam Walsh Protection and Safety Act of 2006," 2006).

Besides actual legislation, the federal government has developed various programs to assist law enforcement and parents with the protection of children online. The Internet Crimes against Children (ICAC) Task Force Program, created by the Office of Juvenile Justice and Delinquency Prevention (OJJDP) in 1998, was developed to help state and local law enforcement construct programs to respond to crimes of online enticement and child pornography. One of the more successful programs, CyberTipline and CyberTipline II (CyberTipline's later enhancement), allows for citizens to report suspicious activity on the Internet, as well as submit unwanted photographs or videos sent to them electronically. In March 2001, the ICAC Task Force reported that because of these citizen reports, more than 550 individuals had been arrested for child sexual exploitation, and 627 search warrants had been served (Medaris & Girouard, 2002).

In May 2006, Attorney General Alberto Gonzales announced the implementation of Project Safe Childhood, a program designed by the Department of Justice to protect children from online abuse. Gonzales encouraged United States Attorneys across the nation to partner with the ICAC task forces and law enforcement officials to develop educational programs that raise awareness of the dangers of online predators and child pornographers. The goal of the program is to increase public awareness and education of Internet dangers so that residents can protect themselves (United States Department of Justice, 2006).

PROTECTIVE MEASURES THROUGH TECHNOLOGY

Many of the statutes passed to criminalize certain materials and activities deemed harmful to minors have been challenged and overturned based on their restriction of free speech that is provided for by the First Amendment. Courts often have suggested the use of filtering and blocking software as an alternative to legislation, which are assumed to be equally effective, but less restrictive (Volokh, 1997). The Clinton administration also endorsed the use of this software, stating that it would do a better job of protecting children from harm on the Internet than any statute (Clinton, 1997). Further federal support came from the chairman of the Federal Communications Commission, William Kennard (1999), who compared the unchaperoned use of the Internet to allowing a child to explore a large city without assistance. Former Vice President Gore (1999) also put forth filtering software as the best tool parents could use to protect their children.

Filtering and blocking software potentially can serve two functions: 1) filtering the receipt of messages, text, or pictures containing certain language, and 2) blocking access to certain sites. According to The Guide (2001), these functions can be further characterized into five different types of software for filtering and blocking certain materials: time-limiting, filtering and blocking, outgoing content blocking, kid-oriented search engines, and monitoring tool. Most family-based Internet safety recommendations endorse the use of filtering and blocking software. However, regulatory advocates have produced studies noting limitations with their use. A study by *Consumer Reports* evaluated six of the mainstream filtering programs and found that all but one, America Online Young Teen Control, blocked at most 20% of sites containing restricted material ("Digital chaperones for kids," 2001). Furthermore, the six programs also blocked a wide range of legitimate content.

Hunter (2000) tested the effectiveness of four popular filtering and blocking software programs: CYBERsitter, Cyber Patrol, Net Nanny, and Surf Watch. He tested the abilities of the programs to block objectionable material, as well as permit non-objectionable material. A website was ranked objectionable or non-objectionable based on its Recreational Software Advisory Council Internet (RSACi) rating, which has five levels of severity based on the content of language, nudity, sex and violence on the website. Websites with a score of two to five were deemed objectionable, and anything below a score of two was non-objectionable. Hunter used three different samples to evaluate the software. The first sample was a set of 50 randomly selected web sites; the second sample was composed of web sites found from a set of 50 popular search terms (i.e., MP3, sex, and Yahoo); and the third sample entailed 100 purposively selected web sites (Hunter, 2000).

Of the four software programs, CYBERsitter was found to do the best job of blocking objectionable material (69%), with Cyber Patrol coming in second with 56% blockage. Surf Watch only blocked 44% of the objectionable material, compared to the 95% blockage claimed in its advertising literature. Finally, Net Nanny did the worst job of blocking material, with only 17% blockage. In regards to blockage of non-objectionable material, CYBERsitter blocked the highest amount with 15%. The other three software packages blocked an average of 6% (Hunter, 2000).

In addition, the second Youth Internet Safety Survey (administered from March 2005 to June 2005), which was discussed previously, found a large majority of parents (84%) to believe that adults should be concerned about their child's exposure to sexual material on the Internet. However, only 33% of these parents had used filtering software in the past year. Five percent of the parents reported using filtering software at one time, but had ceased use by the time of the interview. Parents who reported low trust of their child regarding Internet responsibility, and those who were aware of what their children were doing online, were more likely to utilize the software (Mitchell et al., 2005).

More recent empirical studies examined the effect of different forms of protective measures on adolescent online victimization. Fleming, Greentree, Cocotti-Muller, Elias and Morrison (2006) and Marcum (in-press) found that the installation of filtering and blocking software had no affect on their exposure to inappropriate materials and behaviors and online victimization. Lwin, Stanaland and Miyazaki (2008) further explored protective measures through a quasi-experimental study of 10 to 17 year olds in regard to their experiences with Internet monitoring and mediation by parents. They found that active Internet behavior monitoring by parents decreased the likelihood of participation in risky behaviors online, as well as exposure to inappropriate materials. However, Lwin et al. (2008) noted that the effectiveness of active monitoring decreased the older the adolescent became, which may be a foreshadowing of the results found in the current study considering the age of the sample.

This research indicates that parental concern about online dangers is high, but this concern does not appear to match their actions. Other research also has found that many families, despite their awareness of online dangers, do not use protective software (Finkelhor, Mitchell, & Wolak, 2000; Lebo, 2000). Mitchell et al. (2005) hypothesized several reasons for this lack of use. First, parents may prefer a more active monitoring method rather than relying on computer assistance. Second, parents may be skeptical of the effectiveness of the software, as suggested in the Consumer Reports (2001) findings. Finally, a divide of knowledge between parents and children could affect adoption of the software. Children are generally more computer-savvy than their parents; a lack of awareness about the uses and availability of material on the Internet could cause a lack of preventative measures (Gallo, 1998; The Henry Kaiser Family Foundation, 2000).

SUMMARY

In recent years, there have been a number of studies that have provided descriptive statistics regarding adolescent Internet use, frequency of harassment, and online victimization. This literature has demonstrated that youth in America are using the Internet at a higher frequency (Addison, 2001; Izenberg & Lieberman, 1998; Lenhart et al., 2001; Nie & Ebring, 2000; Rainie, 2006; United States Department of Commerce, 2002), with several modes of computer-mediated communication being employed. However, along with this increased use, studies have shown that youth also are becoming victims of different types of online victimization at a higher rate (Mitchell et al., 2003; (Mitchell et al., 2007; O'Connell et al., 2002; Quayle & Taylor, 2003; Sanger et al., 2004; Wolak et al., 2002; Wolak et al., 2003; Wolak et al., 2004; Wolak et al., 2006).

Relatively few studies have been initiated to examine adult predators who often participate in the online victimization of youth (Armagh, 1996; Beebe et al., 1998; Danet, 1998; Dean, 2006; Durkin, 1997; Durkin & Bryant, 1999; Henderson, 2005; Jones, 1999; Lamb, 1998; McFarlane, Bull, & Rietmeijer, 2000; Quayle & Taylor, 2003; Wolak et al., 2002). Although parents and children are aware of these predators, whether from media reports or actual experience with victimization, they appear to be less than fully aware of the protections available, perhaps due to a lack of information. While government legislation and private entities have made attempts to protect youth from these predators, the problem continues to exist in cyberspace (Fitzpatrick, 2006; Gallo, 1998; Henderson, 2005; Hunter, 2000; Kendall, 1998; McCabe, 2000; Medaris & Girouard, 2002; Mitchell et al., 2005; Mota, 2002; Simon, 1999; Volokh, 1997).

Upon examination of the literature, it can be seen that there is a lack of empirical research that tests a theoretical explanation of online victimization and harassment. In other words, there are few studies to date that present evidence of the relationship between adolescents' online use and their victimization. Assumptions only can be made based on the descriptive data available, greatly limiting knowledge in this area. The current study will seek to fill this gap in the literature by considering an explanation of online victimization based on Routine Activities Theory.

As discussed in the next chapter, Routine Activities Theory often is used to explain various forms of victimization. This theory states that there are three components necessary in a situation in order for a crime to occur: a motivated offender, a suitable target, and a lack of guardianship (Cohen & Felson, 1979; Cohen & Felson, 1981; Felson, 1986; Felson, 1987). This study will survey adolescents (suitable targets) regarding their use of online mediums of communication (possible lack of guardianship and exposure to motivated offenders), as well as their experiences with online victimization and offending. Because of the lack of this type of theoretical research in the current literature, it is expected that this study will provide a better understanding of the phenomenon of adolescent online victimization and allow for larger strides to be made in protecting youth online.

CHAPTER 3:
The Prevalence of Routine Activities Theory

Society and its activity patterns are in a constant state of transformation (Madriz, 1996), especially with the development of new technology. For example, the daily and routine activities of children have evolved from bicycles and dolls to video games and the Internet. Rainie (2006) reported that 87% of youth currently are using the Internet, and that number likely will continue to grow. However, as innovative technologies emerge, new methods of victimization also accompany these developments (Mitchell et al., 2003; O'Connell et al., 2002; Sanger et al., 2004; Wolak et al., 2004; Wolak et al., 2006).

Early tests of Routine Activities Theory, which often is used to examine different types of victimization, focused on the importance of the environment as a vital component of interaction between criminal offenders and victims (Cohen & Felson, 1979). This is particularly relevant to the current research, as the environment, cyberspace, is a necessary factor that must be present in order to both participate in online activities and become a victim of harassment or other online crime. Cyberspace, which thrives on the possibilities of the unknown, also provides the opportunity for engaging in activities without the presence of a capable guardian. This is true for both the offender and victim, as both parties potentially can participate in deviant behaviors without much guardianship being present (Beebe et al., 1998; Danet, 1998; Jones, 1999). According to Felson (1987), a lack of behavioral controls encourages willingness to participate in criminal activity, and motivated offenders will place themselves in areas that have an abundance of suitable targets. For example, youth-oriented chat rooms, instant messaging services, and social networking web sites provide a plethora of opportunities for motivated adult predators. As stated by Felson (1987), it is comparable to "how lion look for deer near their watering hole" (p. 912).

Roncek and Maier (1991) suggested that Routine Activities Theory is excellent for use in the examination of predatory or exploitative crimes, which is precisely the type of deviant behavior examined in this study. Before applying the theory in the context of this research, it is important to examine the theory itself, and how it has been utilized in the past to explain different types of victimization. This chapter first will provide a discussion of the historical development of the theory into its current state. Next, a review and critique of past empirical tests of the theory, based on different types of victimization, will be presented. Finally, after reviewing the theory and its various applications, the three main components of Routine Activities Theory will be applied to the problem of adolescent online victimization.

ROUTINE ACTIVITIES THEORY

The work of Cohen and Felson (1979) was preceded by the work of Hindelang, Gottfredson, and Garofalo (1978), as well as Amos Hawley (1950). Hindelang et al. (1978) developed what is commonly termed "lifestyle/exposure theory," which was based on correlation between lifestyle choices and victimization. They asserted that the variance in victimization risk is related to differences in lifestyle choices. Lifestyle choices encompass the daily activities of a person's life, such as work, school, and extracurricular activities. Choices made by individuals influence their exposure to different persons and places, as well as deviant behaviors, which increases their own risk of victimization (Hindelang et al., 1978). To illustrate, Hindelang and colleagues examined lifestyle choices and demographic characteristics of people more likely to be victims of personal and property crimes. Their findings indicated that younger, unmarried, poor, black, and male individuals are more likely to be victims of personal and property crimes, because they tend to participate in activities away from home, especially after daylight, and they are more likely to be associated with people who are already criminal offenders (Hindelang et al., 1978).

Routine Activities Theory is somewhat similar to lifestyle/ exposure theory (Messner & Tardiff, 1985). According to Brantingham and Brantingham (1981), Cohen and Felson sought to expand and improve upon the work of Hindelang et al. (1978) by incorporating ecological concepts, specifically Hawley's (1950) components of temporal organization: rhythm, tempo, and timing. Rhythm is the regularity with which events occur. Tempo is the number of events that occur per unit of time. Finally, timing is the duration and recurrence of the events. According to Cohen and Felson (1979), the inclusion of these three components improves the explanation of how and why criminal activity is performed.

Cohen and Felson (1979) agreed with the assertion of Hindelang et al. (1978) that the routine activities of certain groups expose them to greater risks of victimization. However, they also argued that changes in our constantly progressing society have provided motivated offenders with more opportunities to commit crime (Cohen & Felson, 1979; Felson, 1994). In general, as a result of changes in people's routine activities (i.e., activities relating to work, home, education, and leisure), the number of available targets has increased, and the presence of capable guardianship has decreased. More specifically, after World War II there was a shift in the amount of routine activities that were performed outside the home. While people are at home, they are guardians of their own domain. However, many modern activities are performed outside the home, and the household is left without a guardian. Cohen and Felson (1979) also asserted that over time, household possessions have become more suitable targets for theft because of easier portability. This claim can be applied to today's technology by comparing music devices from 1980 and 2007. An MP3 player is a valuable commodity and is much easier to steal than a large stereo.

From the initial assertions of Cohen and Felson, and in conjunction with the works of various other scholars, the currently recognized Routine Activities Theory has been formed. This theory states that there are three components necessary in a situation in order for a crime to occur: a suitable target, a lack of a capable guardian, and a motivated offender (Cohen & Cantor, 1980; Cohen & Felson, 1979; Cohen & Felson, 1981; Felson, 1986; Felson, 1987; Hawdon, 1996; Lasley, 1989; Sampson & Wooldredge, 1987). Moreover, crime is not a random occurrence, but rather, follows regular patterns that require these three components.

According to Meier and Miethe (1993), target suitability is based on a person's availability as a victim, as well as his or her attractiveness to the offender. A person who is available for victimization is someone who has not taken certain precautions to protect themselves. For example, leaving a car in the driveway with the keys in the ignition increases victimization risk, because the owner of the car did not take precautions to prevent theft. Attractiveness refers to the appeal of the target based on the value of what the victim has to offer, such as property or sexual favors. With these two aspects present, a person becomes a more likely and suitable candidate for criminal victimization.

The second component necessary for a crime to occur, according to Cohen and Felson (1979), is a lack of capable guardianship. Guardianship is the ability of persons and objects to prevent a crime from occurring (Garofalo & Clark, 1992; Meier & Miethe, 1993; Tseloni et al., 2004) and can take two forms: social and physical. Social guardianship can exist through such factors as household composition, lifestyle, marital status, and employment type. For example, working full-time may ensure that a person will be away from the home roughly 8 hours a day, therefore leaving the home unattended. In contrast, physical guardianship refers to self-protective measures taken by a person, such as burglar alarms and outside lighting. These types of measures have been supported as a method of decreasing burglary rates in neighborhoods (Miethe & McDowall, 1993; Miethe & Meier, 1990; Tseloni et al., 2004).

A motivated offender is a person who is willing to commit a crime when opportunities are presented through the presence and absence of the other two components (Cohen & Felson, 1979; Mustaine & Tewksbury, 2002). In other words, the theory asserts that if a motivated offender is presented with a suitable target that is not properly guarded against victimization, a crime is likely to occur. Schwartz and Pitts (1995) also suggested that offenders are motivated to commit crime because of the support of society to continue the behavior. For example, a person who burglarized an automobile for a compact disc player may be likely to continue this behavior because of a lack of punishment received and the encouragement provided by friends to participate in this lucrative behavior. In summary, situations conducive to crime are those that provide the opportunity to inspired people to take advantage of persons or property left unguarded.

An examination of these three components demonstrates that in general, individuals who enter unsafe environments and participate in risky activities are more likely to become victims of crime. For example, a younger woman who walks home through a dangerous neighborhood every evening at 10:00 p.m. after she gets off work has drastically increased her likelihood of victimization (e.g., rape, assault, murder) based on her routine activities. However, an absence of one or more of the three components of Routine Activities Theory will significantly reduce the potential for victimization. In the previous example, taking public transportation or changing the work shift to daytime could reduce the likelihood of victimization.

Routine Activities Theory also challenges the "pestilence fallacy" (Felson, 1994), which is the assumption that crime is a result of the occurrence of other negative events. The theory insists that crime is actually a result of the socio-structural occurrences of people's everyday activities. For example, Cohen and Felson (1979) originally stated that the rate of direct-contact predatory crime is dependent on the amount of opportunities available to commit crime in everyday situations. More recently, factors such as social structure and demographic characteristics have been found to influence daily routines, which in turn influence exposure to motivated offenders (Gaetz, 2004; Miethe, Stafford, & Long, 1987; Sampson & Wooldredge, 1987). Overall, "the convergence in time and space of suitable targets and the absence of capable guardians can lead to large increases in crime rates" (Cohen & Felson, 1979, 604).

Although Routine Activities Theory could be considered a fairly recent development in theoretical criminology, its evolution is a result of the work of several renowned scholars over time (Cohen & Felson, 1979; Hawley, 1950; Hindelang et al., 1978). The theory provides a logical explanation for victimization by suggesting that the social context of victimization is a key factor in the risk of being victimized (Cohen & Felson, 1979; Collins, Cox, & Langan, 1987; Lynch, 1987; Woolredge, Cullen, & Latessa, 1992). Although an individual may be motivated to commit an offense, a suitable opportunity must be present, or it is unlikely a crime will occur. The following pages will demonstrate how this contemporary theory has been empirically tested and supported on numerous occasions, which in turn provides good reason for its use in the present study.

EMPIRICAL STUDIES

A key aspect of Routine Activities Theory is that crime is not a random occurrence, but rather is a result of regular activity patterns converging in time and space (Mustaine & Tewksbury, 2002). Certain lifestyles and activities enhance victimization risk, based on increased involvement with dangerous persons and places. Routine Activities Theory has been tested and supported by several studies that have explained victimization on the macro-level (Cao & Maume, 1993; Cook, 1987; LaGrange, 1999; Roncek & Bell, 1981; Roncek & Maier, 1991; Sampson, 1987; Tseloni, Wittebrod, Farrell, & Pease, 2004). A more voluminous amount of supporting literature is based on micro-level studies, which examine individual offending behavior (Bernburg & Thorlindsson, 2001; Felson, 1986; Horney et al., 1995; Schreck & Fisher, 2004; Sasse, 2005), personal and property crime victimization (Arnold et al., 2005; Cohen & Cantor, 1980; Cohen et al., 1981; Collins, Cox, & Langan, 1987; Gaetz, 2004; LaGrange, 1994; Lasley, 1989; Lynch, 1987; Moriarty & Williams, 1996; Mustaine & Tewksbury, 1999; Schreck & Fisher, 2004; Spano & Nagy, 2005; Tewksbury & Mustaine, 2000; Woolredge et al., 1992), domain-specific models (Ehrhardt-Mustaine & Tewksbury, 1997; Garofalo et al., 1987; Lynch, 1987; Madriz, 1996; Wang, 2002; Wooldredge et al., 1992), and feminist interpretations (Armstrong & Griffin, 2007; Mustaine & Tewksbury, 2002; Schwartz & Pitts, 1995; Schwartz et al., 2001). This section will discuss the findings from these various studies, as well as provide an assessment of the criticisms of Routine Activities Theory revealed through its empirical tests.

Macro-Level Studies

As stated above, the theory has been used to explain victimization on a macro-level. In general, macro-level research seeks to explain variation in crime rates, such as differences in crime rates between counties, areas within a county, or neighborhoods (Cohen & Felson, 1979). Sherman, Gartin, and Buerger (1989), as well as Tita and Griffiths (2005), have argued that there has been an abundance of use of individual-level data, but not enough utilization of spatial data. The supporting macro-level studies that are available compare crime data for neighborhoods (LaGrange, 1999; Roncek & Bell, 1981; Roncek & Maier, 1991), cites (Cao & Maume, 1993; Cook, 1987; Sampson, 1987), and even countries (Tseloni, Wittebrod, Farrell, & Pease, 2004), to better understand the effect of routine activities in these particular areas.

Neighborhoods
A benefit of conducting macro-level research is the ability to compare crime rates in areas that share a common link, whether it is location or organizational structure. Roncek and Maier (1991), building off the previous work of Roncek and Bell (1981), examined the effect of taverns and cocktail lounges on the crime rate in Cleveland's residential city blocks between 1979 and 1981. The researchers stated there were several reasons that areas with such businesses should experience higher crime rates, consistent with Routine Activities Theory. First, cash would be present on the patrons, as well as in the businesses, making them suitable targets for crime. Second, routine activities in a tavern (e.g., drinking) takes persons away from their intimate handlers and weakens social controls (Hirschi, 1969), in turn influencing them to participate in, or be victims of, deviant activities. Finally, Roncek and Maier (1991) found persons inhabiting these types of businesses may not be residents of the area. Changes in the amount and types of people in an area could decrease guardianship performed by others, as compared to if they were in a less busy residential area.

Data were collected from the Cleveland Police Department on the numbers and types of crimes (i.e., violent and property crimes) reported in 4,396 residential blocks. The average number of residents in each block was 129. There were 547 taverns and 37 cocktail lounges present in the residential blocks at the time of data collection. Multiple regression models found that in all residential blocks, the number of taverns and lounges had a significant effect on the crime rate. Tobit analyses further revealed that the number of taverns and lounges was associated with a higher probability of crime in an area deemed as safe, as well as on blocks that already had more crime (Roncek & Maier, 1991).

LaGrange (1999) also examined the effect of certain factors in residential areas and neighborhoods. She examined the distribution of minor property crimes in a Canadian city during a one-year period in relation to three predictors: neighborhood demographics; proximity of shopping malls; and proximity of public and Catholic junior and senior high schools. The study hypothesized that physical features of an environment with malls and school were expected to produce an increased rate of minor crimes. Shopping malls and schools hinder effective guardianship, because of the difficulty in determining legitimate and illegitimate patrons and the large amount of traffic, along with the considerable amount of young people loitering in the area (LaGrange, 1999).

Records were obtained on all reported vandalism and damage that occurred in the year 1992 from the local police service (n = 13,131), transit department (n = 1,325), and Department of Parks and Recreation (n = 393). Population and housing characteristics were acquired from the 1992 Municipal Census of the city. Ordinary least squares (OLS) regression determined that the presence of high schools and shopping malls within an area were significant predictors of an increase in mischievous events. According to LaGrange (1999), her contribution to the literature provided support for Routine Activities Theory, by demonstrating how a lack of guardianship in an area causes significant increases in criminal activity.

Cities
As mentioned previously, macro-level studies can be used to compare structural aspects of different areas, such as cities or towns. Cao and Maume (1993) performed one of the more noted macro-level studies by examining structural factors that cause robbery occurrences to vary depending on the urban environment. They argued that robbery is a unique violent crime, because it disproportionately occurs in cities and is generally a crime between strangers (Cook, 1987). For further support, they noted that Sampson (1987) proposed that features of an urban area affect the organization of daily activities, which affects the chances an offender and victim will cross paths.

Cao and Maume (1993) selected 296 standard metropolitan statistical areas (SMSAs), which provided continuity with previous related research (Messner & Blau, 1987). The average robbery rate from 1980 to 1982, according to the Federal Bureau of Investigation's Uniform Crime Report, was used as the dependent variable. Urbanization, inequality, and lifestyle indexes were constructed as the independent variables. In regard to control variables, the regional location of the SMSA was accounted for depending on location (coded as 1 if located in the original Confederate states and 0 if located otherwise). An ordinary least squares analysis indicated that urbanization had a direct positive effect on robbery. Large, urbanized areas have residents who are less likely to own a car or home, and people's daily routines are dependent on public transportation. Based on this, people are more likely to cross paths as strangers and motivated offenders, without guardianship, thereby leading to robbery victimization. In summary, urbanism is a strong determinant of a risky routine lifestyle.

Countries

The recent research of Tseloni et al. (2004) examined factors related to burglary occurrences in three countries, based on three victimization surveys taken at approximately the same time: the 1994 British Crime Survey (N = 12,845), the 1994 National Crime Victimization Survey (N = 72,412), and the 1993 Netherlands Police Monitor (N = 39,849). The British Crime Survey is a biannual survey given to adults over 16 in randomly selected households, through the use of a nationwide address list. The sample for the annual National Crime Victimization Survey is acquired in a similar manager as the British Crime Survey, with the same respondent ages. The Police Monitor, also completed biannually, is a representative survey of households produced through telephone contact.

Although there were differences in the frequencies across data sets, several similar lifestyle characteristics were shown to be significant indicators of victimization. Negative binomial regression models revealed that proximity to potential offenders (motivated offenders), a city size of 250,000 persons or more (availability of suitable targets), and lone parenthood (lack of guardianship) provided the strongest impact on potential victimization in all three areas. The results of this study provided additional support for Routine Activities Theory (Tseloni, Wittebrod, Farrell, & Pease, 2004).

Summary

Although not as abundant as micro-level studies, macro-level studies have demonstrated their usefulness for testing Routine Activities Theory. Lack of guardianship has been shown to have a significant effect on crime rates in neighborhoods, especially in areas with large amounts of traffic from non-residents that have no ties to the area

(LaGrange, 1999; Roncek & Bell, 1981; Roncek & Maier, 1991). A comparison of crime data between cities also has demonstrated the effect of a lack of guardianship and risky lifestyles on increased victimization (Cao & Maume, 1993; Cook, 1987; Sampson; 1987). Even an examination of countries in different continents revealed support for the theory, by demonstrating how a lack of guardianship, combined with motivated offenders crossing paths with suitable targets, increases the likelihood of victimization (Tseloni et al., 2004).

These types of studies also have been found to have strong predictive capabilities (Kennedy & Forde, 1990), which is important when examining changing crime rates in urban environments. However, macro-level research usually does not measure routine activities directly (Bennett, 1991), instead relying on aggregate measures, which often can lead to overlooking important factors identified through individual-level analysis. With this in mind, in the mid-1990s Osgood, Wilson, O'Malley, Bachman, and Johnston (1996) argued there was a lack of use of individual data to explain victimization, with too many studies relying on the use of aggregate data. The next section will examine studies that did employ individual-level data, rather than aggregate measures.

Micro-Level Studies

Osgood et al. (1996) further asserted that Routine Activities Theory would be useful in explaining individual offending through micro-level research. They suggested a person's increased exposure to easy opportunities for deviant behavior would increase the likelihood of a crime. Furthermore, they argued micro-level research allows for a better explanation of this phenomenon, while holding constant alternative causal factors. Several previous studies had, in fact, examined the routine activities of individuals and provided support for the hypothesis that crime is more likely to occur when these activities increase the likelihood of the union of the three key theoretical components: a suitable target, a lack of guardianship, and a motivated offender (Forde & Kennedy. 1997; Hawdon, 1996; Lasley, 1989; Meier & Meithe, 1993; Miethe et al., 1987; Osgood et al., 1996; Schreck & Fisher, 2004).

The following micro-level research examines individual offending and victimization in different forms. These studies focus on offending behavior, personal and property crime victimization, and domain-specific models. Also, an examination of Routine Activities Theory from the feminist perspective will be presented, as well as miscellaneous studies that examined the general risk of victimization. The review of this literature will demonstrate the value of using Routine Activities Theory to explain victimization, based on the continued support found in these studies.

Offending behavior

The majority of the research on this theory examines the behaviors and experiences of the victim. However, some research has examined the behaviors of offenders and how the components of Routine Activities Theory affect their choice to commit crime. One cross-sectional study focused on the third component of Routine Activities Theory, lack of guardianship, by examining survey data from Icelandic adolescents (Bernburg & Thorlindsson, 2001). All students in the compulsory ninth and tenth grades of Icelandic secondary schools were administered anonymous questionnaires in March 1997. Two versions of the questionnaire were randomly distributed to the respondents, containing the same core questions, but with greater emphasis on different topics depending on the version of the questionnaire. A nationally representative sub-sample (n = 3,260) was taken from the entire sample of returned questionnaires (N = 7,785) for this particular study.

Respondents were questioned about their behavior and activities during periods of time without adult supervision. Ordinary least squares regression was used to examine relationships between routine activities indicators and deviant behavior. Unstructured peer interaction was found to have a significant positive effect on property offending. Moreover, the data also indicated that unstructured peer interaction had a weaker effect on deviant behavior when the level of social bonding increased; therefore, this study provided only moderate support for Routine Activities Theory regarding the tenet of lack of guardianship.

An offending behavior receiving much scrutiny is sexual assault. One study on this particular type of behavior came from Sasse (2005), who examined the motivating variables of sex offenders. He asserted that motivation to commit these types of offenses is acted upon not only because the motivation is available, but also because it originates from histories of past abuse, low self-esteem, and other psychological variables. This study hypothesized that according to Routine Activities

Theory, these variables should influence behavior based on the proximity of the offender to the victim, which influences the level of motivation.

Data were collected from 163 men and women in three counties in a Midwest community-based sex offender program. Ninety-six percent of the offenders were male. All respondents participated in the program between 1982 and 2000. At intake, the respondents were subjected to a variety of assessments that included questions on personal demographics, offense characteristics, victim relationship, and abuse histories (Sasse, 2005).

Results from a logistic regression model found general support for the theory. Home offenders were more likely to be abused and have younger victims compared to their community correspondents. These types of offenders have an advantage, as they have more time to groom their victims and influence them to conceal the behavior. Community offenders were younger, with older victims, and were more likely to have used alcohol before the offense. Again, support was found for the theory, as the targets for these offenders were at the age that roaming the community unsupervised is allowed.

Another examination of sexually deviant behavior more recently came from Jackson, Gililand, and Veneziano (2006). Their purpose was twofold: 1) to examine the prior and current deviance of male college students in relationship with athletic participation, fraternity participation, and opportunity, specifically in the prediction of sexual aggression; and 2) to assess the relationship between prior delinquency and sexual aggression. A sample of 304 college men involved in various contact and non-contact sports, as well as fraternity and non-fraternity members, was surveyed on previous and current behaviors. The average age of the respondents was 20 years old.

Analytic moment structures (AMOS) analysis indicated that prior deviance had a positive and significant effect on levels of sexually aggressive behavior. College-level deviance was also a significant indicator of sexually aggressive behavior. This behavior also was predicted by opportunity to commit the act. In other words, there is a higher likelihood an offender will be sexually aggressive if the opportunity to potentially commit the offense is present (Jackson, Gililand, & Veneziano, 2006).

Personal and property crime victimization

Routine Activities Theory has been used frequently in the past to explore personal and property crime victimization (Arnold, Keane, & Baron, 2005; Cohen & Cantor, 1980; Collins, Cox, & Langan, 1987; Gaetz, 2004; Lasley, 1989; Moriarty & Williams, 1996; Mustaine & Tewksbury, 1999; Schreck & Fisher, 2004; Spano & Nagy, 2005; Tewksbury & Mustaine, 2000). For example, early research by Cohen and Cantor (1980) and Cohen, Cantor, & Kluegel (1981) analyzed data from the National Crime Survey to consider the relationships between lifestyle and demographic characteristics and larceny victimization. The findings from both studies supported Routine Activities Theory. Risk of victimization was found to be higher for younger persons who lived alone and were unemployed (Cohen & Cantor, 1980; Cohen et al., 1981).

The theory suggests that the social context of criminal victimization is a central factor in the risk of victimization (Cohen & Felson, 1979; Lynch, 1987; Woolredge, Cullen, & Latessa, 1992). In other words, even though a person may be motivated to commit an offense, the lack of a suitable target and the presence of proper guardianship will make a crime unlikely to occur. To illustrate, college students are often victims of property crime, especially vandalism (Sloan, 1994). Tewksbury and Mustaine (2000) tested the hypothesis that a target increases its chances of becoming a victim of vandalism based on the movement of property in public domains. Data for this study were taken from surveys given to 1,513 university students enrolled in introductory-level criminal justice and sociology courses at 9 universities and colleges in 8 different states. The majority of respondents were full-time, heterosexual students under the age of 21. The survey instrument questioned respondents on daily routines, illegal activities, substance use, and community structural variables.

Logistic regression analysis indicated that a person's drug and alcohol use, as well as his or her leisure activities, were not significant predictors of vandalism victimization. Home security measures also had no effect on the victimization of these college students. However, employment status and neighborhood characteristics were found to be significant predictors. Those who were employed had a 0.35 decrease in the log odds of being victims of vandalism. Tewksbury and Mustaine (2000) reasoned that this is because those who are unemployed leave their property unsupervised for long periods of time, possibly because they are in class or looking for work, and therefore expose it to potential offenders. In regard to neighborhood characteristics, persons who lived near a park had a 0.58 increase in the log odds of victimization, seemingly because parks contain a large number of unsupervised persons (i.e., potential offenders) who are exposed to suitable targets to victimize.

Gaetz (2004) further investigated the personal and property victimization of homeless youth through the application of Routine Activities Theory. His hypothesis was that homeless youth, based on their situational factors, would be more likely to be victimized than youth who had constant shelter. The study involved surveys and interviews with 208 homeless youth in Toronto, Canada, as well as the examination of data from Canada's General Social Survey (CGSS) (Statistics Canada, 1999). Research on the homeless youth was conducted at 8 service agencies during the fall of 2001, involving young people between the ages of 15 and 24 who had been homeless during the previous year. The CGSS was conducted in 1999 through telephone interviews of 26,000 Canadian residents ages 15 and older. Homeless people were excluded from those interviews, since they could not be contacted by home telephone.

According to Gaetz (2004), the results of the CGSS revealed that half of the reported victimizations involved personal crimes and 35% involved household crimes, such as vandalism and motor vehicle theft. Young people ages 15 to 24 who participated in the GSS experienced higher levels of victimization compared to older adults; 18% were victimized more than once (Statistics Canada, 1999). On the other hand, the homeless youth survey indicated that homeless youth were at even greater risk for victimization. Approximately 82% of these youth had been victims of crime, and 79% reported being victimized more than once. A majority of the victimization involved personal crimes, such as assault (Gaetz, 2004).

Situational factors, such as personal activities and the places frequented, are a central feature of Routine Activities Theory. In Gaetz's (2004) study, the theory was extended by applying the concept of social exclusion, which explores the restriction of a person's access to institutions and practices based on their economic, social, and political conditions (Mandaipour, 1998). Homeless youth are restricted in the most obvious resource for enhancing safety: constant shelter. According to Gaetz (2004), data from this study support Routine Activities Theory, as it was shown that homeless youth experience twice the amount of victimization as youth who have constant shelter. They are therefore unable to avoid dangerous persons and situations because of their inability to obtain shelter, and in turn are restricted in the manner in which they can protect themselves.

In general, family and peer connections are life mechanisms that allow for exposure to motivated offenders, as well as the presentation of suitable targets, for young people. The level of social bonds a person develops has been shown to contribute to routine activities that affect the risk of victimization (Felson, 1986; Horney, Osgood, & Marshall, 1995). A recent study by Schreck and Fisher (2004) examined the hypothesis that strong attachment to family encourages a better quality of guardianship by parents, in turn making children less attractive targets by limiting their contact with motivated offenders, because they are spending more time at home (Felson, 1986; Shreck, Wright, & Miller, 2002). Data were obtained from the first wave of the National Longitudinal Study of Adolescent Health, which was conducted between September 1994 and December 1995. Surveys were administered to a nationally representative sample of students in grades 7 through 12 (N = 3,500), followed by in-home interviews of the students 1, 2, and 6 years later. The purpose of the study was to learn about the effect of social environment on adolescent health. Violent victimization was the dependent variable assessed through a Poisson regression model, while testing for the effects of family context, peer delinquency, peer context, lifestyles, and demographic variables (Schreck & Fisher, 2004).

The results of the study identified several variables that influenced violent victimization in adolescents. First, minorities were more likely to experience victimization than Caucasian youth. Second, family climate and parental feelings towards children were the strongest family-related predictors of victimization. Adolescents living in homes with a loving environment appear to be at less risk for victimization, because they spend more time in the home rather than unsafe environments that make them suitable targets. Finally, peer delinquency and activities with friends were the strongest peer-context variables that influenced victimization, and teenagers who participated in higher levels of delinquency were more likely to be victimized because they were exposed to other motivated offenders. Overall, these results were supportive of the main hypothesis of the study (Schreck & Fisher, 2004).

Although there is a voluminous amount of research investigating adolescent violence, Spano and Nagy (2005) argued that little research has been conducted examining determinants of rural violence using targeted samples of rural youth. Using data from the Alabama Adolescent Survey, administered in 2001, Routine Activities Theory was utilized to explain assault and robbery victimization among rural adolescents, and social isolation was assessed as a risk or protective factor. Sixteen school districts in six poor and rural counties were surveyed; no school had more than 200 students in each grade level. Only grades nine and ten were surveyed, and 90% of the students present on the day the survey was distributed actually completed the survey.

Multivariate logistic regression was used to examine the relationships between assault and robbery victimization and the five types of independent variables: demographics, measures of deviant lifestyle, teasing, social isolation, and social guardianship. The statistical analysis indicated that deviant lifestyle indicators and teasing increased the probability of assault victimization. Gender, level of teasing, and social isolation had a significant effect on robbery victimization. Surprisingly, black adolescents were less likely than white adolescents to become victims of assault and robbery, which deviated from the national trend. Spano and Nagy (2005) accredited this difference to the possibility that black residents of rural areas may be more watchful and have stronger friendship connections, and that more blacks go to the public school system in the rural portion of Alabama that was surveyed. Based on these findings, the authors suggested that Routine Activities Theory may need to be altered to accommodate community influences in rural areas that affect victimization.

A more recent study by Arnold, Keane, and Baron (2005) examined how causal factors affect the risk of personal and property victimization for the general population. Rather than concentrating on offender motivation, which Cohen and Felson (1979) asserted was always present, the researchers focused on suitable targets and lack of guardianship in the equation. Data were collected in the Canadian General Social Survey of 1988, through the random digit dialing technique. The sample (N = 9,551) was composed of respondents in 10 provinces in Canada, who were non-institutionalized and over the age of 15. The participants were questioned on their experiences with various lifestyle aspects and criminal victimization falling into four categories: violence, theft of personal property, household crime, and vehicle theft (Arnold et al., 2005).

Standard logistic regression indicated a few significant predictors of victimization. First, respondents under the age of 24 were more likely to be victimized. Second, the main activities of the respondent (such as drinking and leisure activities) drastically increased a person's likelihood of being a victim of violence or a household crime. Finally, it was discovered that respondents with a high risk of one form of victimization were more likely to have a high risk of another form, due to their lifestyle choices. In other words, people who participate in activities that make them vulnerable as targets, with little guardianship, are likely to be victimized in many forms (Arnold et al., 2005).

In sum, much of the past research performed on Routine Activities Theory has provided support for the theory by exploring several types of property crime, such as vandalism and motor vehicle theft (Cohen & Cantor, 1980; Cohen et al., 1981; Massey et al., 1989; Bennett, 1991; Tewksbury & Mustaine, 2000; Gaetz, 2004). Support for the effect of routine activities on personal crime, such as assault and robbery victimization, has also been revealed in past studies (Arnold et al., 2005; Gaetz, 2004; Spano & Nagy, 2005). In the next section, the idea of situational factors being related to victimization will be considered in more detail.

Domain-Specific Models

Many researchers have used domain-specified models to better explain routine activities inside and outside the home environment (Ehrhardt-Mustaine & Tewksbury, 1997; Garofalo, Siegel, & Laub, 1987; Lynch, 1987; Madriz, 1996; Wooldredge, Cullen, & Latessa, 1992). This approach allows for a closer examination of a specific area of criminal activity, allowing for a more focused look at a particular type of environment. By examining victimization in a certain area rather than in generalities, a better explanation of victimization potentially can be provided (Wooldredge, 1998). Although infrequently used, domain-specific models have been well supported when studying routine activities (Lynch, 1987; Madriz, 1996).

Wooldredge, Cullen, and Latessa (1992) analyzed victimization in the workplace by testing Routine Activities Theory in a specific environment. These researchers stated that the majority of studies testing this theory were not domain-specific when questioning a person on victimization; therefore, it is difficult to conclude a direct relationship exists between routine activities and the likelihood of victimization (Lynch, 1987). Full-time faculty members at the University of Cincinnati, working on the west campus between September 1, 1989, and December 31, 1990, were polled through a self-report questionnaire on their personal and property crime victimization experiences on campus. The faculty who responded (N = 422) also answered questions about their demographic characteristics, work environment, daily activities, and feelings on campus safety.

Logistic regression was used to analyze the causes and types of victimization for these faculty members. One hundred sixteen faculty members reported being victims of property crimes, and 21 were victims of personal crimes. Demographic characteristics were not found to be significant predictors of victimization. However, variables representing extent of exposure to motivated offenders were significant predictors of property crime victimization. Level of guardianship variables, such as the amount of time faculty spent on campus after hours and faculty whose offices were not located within shouting distance of other faculty, were also significantly related to property crime victimization (Wooldredge, Cullen, & Latessa, 1992). However, none of these guardianship variables were predictors of personal crime victimization, and target attractiveness, which was measured by the location of faculty offices on campus and possessions available in the office, was found to be an insignificant predictor of both types of crimes. The authors noted that these findings were not surprising, as faculty members generally have control over who enters their offices, as well as their possessions (Wooldredge, Cullen, & Latessa, 1992).

Another study initiated by Wang (2002) examined criminal activity in the specific domain of banks. He asserted that bank robberies are not a random event, but instead are based on determinative social circumstances that allow banks to be suitable targets. The focus of this study was on the causal factors associated with bank robberies by Asian offenders. Six police incident reports from six different robberies were culled for data, along with interviews of investigators and bank employees. These data indicated that the perpetrators in each robbery were Asian gang members between the ages of 18 and 25.

Wang considered all three components of Routine Activities Theory when collecting the data, and he found support for his hypothesis that these components were important during the criminal event. First, he stated motivation was clearly present based on the careful planning and different approaches used by the gang members to rob the bank. Second, each bank was an apparent suitable target, having excessive amounts of cash and being located close to a major highway for easy escape. Finally, each bank lacked in guardianship, based on inadequate security measures at the time of the robbery. Either the banks were robbed early in the morning, when there was little to no security, or the security guards were unarmed.

Overall, domain-specific models of victimization allow for a closer examination of the specific area of the criminal activity, allowing for a narrow scope of inquiry. They have been used to examine victimization in the home, workplace, and other specific places containing a person's routine activities (Ehrhardt-Mustaine & Tewksbury, 1997; Garofalo, Siegel, & Laub, 1987; Lynch, 1987; Madriz, 1996; Wooldredge, 1998; Wooldredge, Cullen, & Latessa, 1992). Scholars in this area have argued that although the activities of offenders and victims often are discussed, the location of the incident is often not considered. By examining the location of crime, it is felt that crime can be better understood and explained. However, this type of model is fairly new, and in order to accurately test Routine Activities Theory, appropriate measures must be used, or the explanatory power of the model will be lower and the ability to generalize the findings to a larger population will be lessened (Wooldredge, 1998).

Feminist Interpretation of Routine Activities

One of the more interesting adaptations of Routine Activities Theory is the feminist interpretation, made popular by Schwartz and Pitts (1995). They attempted to better explain motivation by describing the reasons males sexually assault women (the suitable targets). They asserted that two factors increase female suitability: 1) women who go out drinking are more likely to be sexually assaulted; and 2) these women are more likely to report they know men that get women drunk in order to sexually assault them. In addition, the presence or absence of capable guardians influences the likelihood of the event (Schwartz & Pitts, 1995). This adaptation of Routine Activities Theory has been tested through data collected from undergraduate males and females, as described below.

Schwartz, DeKeseredy, Tait, and Alvi (2001) investigated the likelihood of sexual assault on a college campus by applying Schwartz and Pitts' (1995) feminist model of Routine Activities Theory. In this particular study, the data were derived from a Canadian sample of community college and university students. Two questionnaires, one for women and one for men, were administered in 95 undergraduate classrooms at 27 universities and 21 colleges. Women were questioned regarding their experiences as a victim of sexual assault, and men were questioned regarding their experiences as an aggressor of sexual assault. The sample (N = 3,142) was composed of 1,835 women and 1,307 men who had ever dated a member of the opposite sex. Female alcohol and drug use and its relationship with sexual assault were assessed through linear analysis. Sexual assault, alcohol and drug use, and male peer support were the key variables that were examined through dichotomous logistic regression (Schwartz et al., 2001).

There was a drastic difference in victimization rates found for women who used drugs (89.3%) and those who did not (41.2%). With regard to alcohol use and victimization, 41.3% of light drinkers were sexually assaulted, compared to 55.2% of heavy drinkers. Bivariate analysis further revealed a significant relationship between sexual assault and level of alcohol and drug use for both men and women. The logistic regression model then indicated the following factors significantly increased the odds that a male will commit an act of sexual aggression: drinking two or more times a week; male peer support for emotional violence; and male peer support for physical and sexual violence. Furthermore, a male with all three of these characteristics was almost 10 times as likely to commit an act of sexual aggression compared to a male with none of the characteristics (Schwartz et al., 2001).

This study provided support for feminist Routine Activities Theory in several ways. First, the data showed that men who drink heavily are more likely to be motivated offenders, and women who consume large amounts of alcohol are more likely to be suitable targets. Second, male peer support for committing acts of sexual victimization against women can be viewed as a motivator, as well as a component of guardianship. Overall, the data from the study demonstrated that sexual victimization could be attributed to participation in routine activities.

A second study utilizing feminist Routine Activities Theory assessed two items: 1) how women's lifestyles affect their suitability as potential victims of sexual assault and 2) whether the factors that influence sexual assault vary depending on the degree of sexual assault (Mustaine & Tewksbury, 2002). The data were collected from self-report surveys collected during the 1998 fall academic terms from undergraduates (N = 1,196) at 12 southern institutions. Respondents were questioned on their demographics, daily routines, and alcohol and drug use. Of those who responded, the majority was female (55.5%) and white (73.7%).

Based on logistic regression models, guardianship behavior did not influence the likelihood of sexual assault. Also, lifestyle choices relating to alcohol did not have a significant effect in this sample, although they have been found to be significant predictors in other studies (Schwartz & Pitts, 1995; Vogel & Himelein, 1995). However, several significant influences on sexual assault were found in the data. Women who participated in leisure time with friends "doing nothing" had higher odds of being victimized than women who did not participate in that type of activity. Also, females who were members of a higher number of clubs and organizations had higher odds of being victimized compared to others in only a few clubs. Finally, and most significantly, women who bought drugs and women who spent more time in public while using drugs had higher odds of being victimized (Mustaine & Tewksbury, 2002).

The utilization of the feminist perspective of Routine Activities Theory is beneficial for theorists when trying to investigate female victimization. More specifically, Routine Activities Theory has been supported through the examination of female sexual assault (Mustaine & Tewksbury, 2002; Schwartz & Pitts, 1995; Schwartz et al., 2001). However, an obvious limitation to this adaptation is that it disregards the importance of male victimization and experiences. Although useful for particular types of crimes, this limits the utilization of this particular adaptation of the theory.

Summary

The micro-level studies listed above utilized individual-level data, which allows for analysis of factors that specifically apply to individuals, rather than across larger groups. This research provides support for the use of Routine Activities Theory in the current study, as it demonstrates the theory's value when explaining personal victimization. The outcome variables assessed in prior research include offending behavior and personal and property crime victimization. The literature on offending behavior, exploring the experiences of the offender and not the victim, revealed that unstructured peer interaction and a lack of parental supervision and connection reflected low levels of guardianship that were significant predictors of criminal offending (Bernburg & Thorlindsson, 2001; Felson, 1986; Horney et al., 1995; Schreck & Fisher, 2004; Sasse, 2005). Personal and property crime victimization studies also provided strong support for the theory, by demonstrating how a person's routine activities (such as participating in leisure activities away from the home and other lifestyle choices), significantly increase the likelihood of victimization (Arnold et al., 2005; Cohen & Cantor, 1980; Cohen et al., 1981; Collins, Cox, & Langan, 1987; Gaetz, 2004; LaGrange, 1994; Lasley, 1989; Lynch, 1987; Moriarty & Williams, 1996; Mustaine & Tewksbury, 1999; Schreck & Fisher, 2004; Spano & Nagy, 2005; Tewksbury & Mustaine, 2000; Woolredge et al., 1992).

The remaining specific categories of research reviewed in this chapter were domain-specific models and the feminist interpretation of Routine Activities Theory. Domain-specific models have been noted to better assess routine activities in a specific environment (Ehrhardt-Mustaine & Tewksbury, 1997; Garofalo et al., 1987; Lynch, 1987; Madriz, 1996; Wang, 2002; Wooldredge et al., 1992). In the particular studies in this review, extent of exposure to motivated offenders and lack of guardianship were key variables with regard to criminal victimization. The feminist interpretation of the theory has attempted to better explain sexual assault victimization (Schwartz & Pitts, 1995). Current studies reveal that drug and alcohol consumption is a significant predictor of the sexual victimization of females (Mustaine & Tewksbury, 2002; Schwartz et al., 2001).

Support for Routine Activities Theory found through micro-level data has been criticized by some scholars who assert that it does not demonstrate as strong of support as found through macro-level studies (Miethe et al., 1987; Sampson & Wooldredge, 1987). Also, Sherman, Gartin, and Buerger (1989), as well as Tita and Griffiths (2005), have argued there has been an abundance of use of individual-level data, but little utilization of spatial data. Nevertheless, the volume of support found in micro-level studies, in conjunction with the macro-level studies also examined, presents a collaborative and undeniable verification of the value of the theory and its usefulness in explaining various types of victimization.

Summary and Application of Routine Activities Theory to the Current Study

Based on an examination of the relevant literature, Routine Activities Theory has received support on both the macro- and micro-level. Although not as plentiful as micro-level research, macro-level investigations of Routine Activities Theory have revealed empirical support for the components of the theory. In particular, lack of guardianship in areas with large amounts of traffic from non-residents having no ties to the area was shown to produce a significant effect on crime rates in neighborhoods (LaGrange, 1999; Roncek & Bell, 1981; Roncek & Maier, 1991). Moreover, the lack of guardianship and risky lifestyles of city residents have a significant relationship with victimization (Cao & Maume, 1993; Cook, 1987; Forde & Kennedy, 19997; Sampson; 1987). An examination of countries in different continents also revealed support for the theory, by demonstrating how not only a lack of guardianship, but also crossing paths with a motivated offender as a suitable target, increases the likelihood of victimization (Tseloni, Wittebrod, Farrell, & Pease, 2004).

Micro-level studies utilize individual-level data, allowing for analysis of factors that specifically apply to individuals, rather than across large groups. Literature on offending behavior indicates unstructured peer interaction and lack of parental supervision and connection reflected a lack of guardianship that was a significant predictor of criminal offending (Bernburg & Thorlindsson, 2001; Felson, 1986; Horney et al., 1995; Schreck & Fisher, 2004; Sasse,

2005). Personal and property crime victimization studies suggest a person's routine activities, such as participating in leisure activities away from the home and other lifestyle choices, significantly increase the likelihood of victimization (Arnold et al., 2005; Cohen & Cantor, 1980; Cohen et al., 1981; Collins, Cox, & Langan, 1987; Gaetz, 2004; LaGrange, 1994; Lasley, 1989; Lynch, 1987; Moriarty & Williams, 1996; Mustaine & Tewksbury, 1999; Spano & Nagy, 2005; Tewksbury & Mustaine, 2000; Woolredge et al., 1992). Domain-specific models were noted to better explain routine activities in a specific environment (Ehrhardt-Mustaine & Tewksbury, 1997; Garofalo et al., 1987; Lynch, 1987; Madriz, 1996; Wang, 2002; Wooldredge et al., 1992). Finally, current studies reveal that drug and alcohol consumption is a significant predictor of sexual victimization of females (Mustaine & Tewksbury, 2002; Schwartz et al., 2001).

Routine Activities Theory asserts that there are three components necessary in a situation in order for a crime to occur: a motivated offender, a suitable target, and a lack of guardianship (Cohen & Felson, 1979; Cohen & Felson, 1981; Felson, 1986; Felson, 1987). Currently, there is a lack of research that provides an explanatory analysis of online victimization of youth; studies to date most often are limited to descriptive data. Along with a general lack of advanced statistical analysis on this type of victimization, there have not been any recorded applications of Routine Activities Theory to the online victimization of youth. In seeking to make a contribution to the literature, this study will not only attempt to explain online victimization, but will do so using a well-supported theory. As demonstrated in this chapter, Routine Activities Theory has been found to provide an explanatory framework for various types of crime victimization, in turn suggesting its value for explaining online victimization.

This study surveyed adolescents (suitable targets) regarding their use of online mediums of communication (possible lack of guardianship and exposure to motivated offenders), as well as their experiences with online victimization and formation of relationships with online contacts. It was hypothesized that greater exposure to motivated offenders who commit these acts of victimization toward youth, as well as a lack of guardianship experienced by youth while using the Internet, would increase the likelihood of victimization and formation of relationships with online contacts. Because of the lack of this type of theoretical application in the current literature on this topic, it was expected that this study would provide a significant contribution for understanding the phenomenon of adolescent online victimization, and potentially allow for larger strides to be made in protecting youth online.

CHAPTER 4

Investigating Adolescent Online Victimization and Formation of Relationships with Online Contacts

The purpose of this study was to examine Internet use and types of use by freshmen at a mid-sized university in the northeast, in relation to their online victimization and formation of relationships with online contacts. There have been few explanatory studies to date that assess the relationship between adolescents' online use and their victimization and formation of relationships with online contacts. In order to more fully examine this phenomenon, the chosen methodology was developed under the concepts and propositions of Routine Activities Theory, which has been utilized many times in the past to explain various types of victimization. This study employed a survey in an effort to produce a more complete understanding of adolescent Internet use and resulting outcomes.

The survey questioned respondents on past and present Internet use and other relevant variables. Specifically, surveys were administered to newly enrolled freshmen in the spring of 2008, with a focus on their frequency and types of Internet use, development of relationships with online contacts, and experiences with different types of victimization and offending. Respondents were asked about their experiences and behaviors as high school seniors, as well as current experiences and behaviors as college freshmen. The students also answered questions regarding their experiences as both an online victim and a participant in the formation of relationships with online contacts.

This chapter first provides a description and discussion of the methodology employed. The sampling and administration plan for the survey then is presented. Next, a discussion of the analysis plan is provided, and based on the quantitative nature of the study, the statistical analyses are discussed. Finally, due to the use of human subjects for examining potentially sensitive material, the protections that were available to the participants are noted.

SURVEY OF STUDENTS

A survey was appropriate for this study, as the purpose of the survey was "to produce statistics, that is, quantitative or numerical descriptions about some aspects of the study population" (Fowler, Jr., 2002, pp. 1). In general, surveys allow for the collection of data to produce the desired statistics by asking people questions about the topic of interest. According to Bachmann & Schutt (2007), survey research continues to be a popular way to collect information, because of three features. First, surveys are versatile, as there is hardly any topic of interest that social scientists cannot examine through a survey. Second, surveys are efficient, because they can collect data on many variables relatively quickly and in a low cost manner. Finally, surveys are often the only feasible method to develop a representative picture of the attitudes and traits of a large population. When generalizability is the main goal of research, surveys are typically a suitable choice.

RESEARCH DESIGN

The surveys for this study were dispensed and collected through the group administration method, utilizing students at a mid-sized university in the northeast while they were in the classroom setting. According to Fowler, Jr. (2002), this method has several advantages. First, surveys of this type generally are known to have high cooperation rates, which reduce the potential for nonresponse error (Dillman, 2007). Second, the administrator has the opportunity to explain the study and answer questions about the survey, in contrast to mail surveys for which interpretation is completely up to the respondent. Finally, there is a lower cost associated with this method. The only expense was the cost of reproduction of the surveys.

There are two common limitations of group-administered surveys. First, assembling a group is sometimes difficult, typically requiring a captive audience (Bachmann & Schutt, 2007). However, this was not an issue for this particular study, as groups of students already were assembled for regular class periods. In turn, this also minimized the amount of coverage error occurring during survey distribution (Dillman, 2007). The second major concern is that respondents will be coerced to participate, and therefore they will be less likely to answer questions honestly (Bachmann & Schutt, 2007). This issue was addressed by emphasizing to participants that completion was completely voluntary and refusal to participate in the survey would not result in any sort of penalty to persons in the class. Also, because this study examined the Internet use of the research population, the topic may have been of particular interest to the respondents, thereby increasing their willingness to complete the survey.

RESEARCH QUESTIONS AND HYPOTHESES

The frequency of Internet use and online victimization experienced by adolescents and college students has been measured in a number of studies (Beebe et al., 2004; Clemmitt, 2006; Freeman-Longo, 2000; Lenhart et al., 2001; Lorig et al., 2002; Mitchell et al., 2003; Mitchell et al., 2007; O'Connell et al., 2002; PRNewswire, 2006; Reeves, 2000; Sanger et al., 2004; United States Department of Commerce, 2002; Wang & Ross, 2002; Wolak et al., 2002; Wolak et al., 2003; Wolak et al., 2004; Wolak et al., 2006). Only a few of these studies, however, have utilized any type of advanced statistical analysis in an attempt to explain the relationship between different facets of Internet use and victimization, along with the formation of relationships with online contacts. Therefore, this study contributes to the existing body of literature, as data collected through the survey was used to test Routine Activities Theory in the context of adolescent Internet use.

Each general research question listed below was developed through the review of past literature, along with a consideration of the three components of Routine Activities Theory, with each component of the theory defined to suit the purpose of this study. Exposure to motivated offenders was indicated through spending extended amounts of time online and in various locations. Target suitability consisted of actions taken by students that increase their attractiveness for victimization, such as providing personal information to online contacts. Finally, lack of guardianship was a deficiency in protective measures (e.g., filtering and blocking software, monitoring by another person) experienced by adolescents during their online use. Based on these constructs, the following research questions were considered through the use of the survey:

1) Does the amount of time and location spent on the Internet increase the likelihood of online victimization and formation of relationships with online contacts?

2) Does revealing personal information to online contacts increase the likelihood of online victimization and formation of relationships with online contacts?

3) Does a lack of supervision or protective measures increase the likelihood of online victimization and formation of relationships with online contacts?

The following specific hypotheses were formulated based on the above three research questions, a review of the available literature, and the purpose of the study. The null hypotheses (H_o) generally state that there are no significant relationships or significant effects with regard to the variables under analysis. The first alternative hypothesis listed below tested the theoretical construct of exposure to motivated offenders, as derived from Routine Activities Theory:

$H_a(1)$: Adolescents who spend more time on the Internet using modes of computer-mediated communication are more likely to be victimized online and form relationships with online contacts.

The following alternative hypotheses were tested for the theoretical construct of target suitability:

$H_a(2)$: Adolescents who provide personal information to online contacts are more likely to be victimized online.

$H_a(3)$: Adolescents who do not privatize their social networking websites for viewing by only approved online contacts are more likely to be victimized online and form relationships with online contacts.

$H_a(4)$: Adolescents who provide personal information to online contacts are more likely to form offline relationships with contacted individuals.

The remaining alternative hypotheses were tested for the theoretical construct of lack of capable guardianship:

H$_a$(5): Adolescents who utilize protective software are less likely to be victimized online.

H$_a$(6): Adolescents who have restricted use of the Internet are less likely to be victimized online and form relationships with online contacts.

H$_a$(7): Adolescents who are monitored while using the Internet are less likely to be victimized online and form relationships with online contacts.

SAMPLE

The population for the present research included all freshmen students enrolled at a mid-sized university in the northeast during the spring 2008 academic term. The ideal age group for sampling would be youth between the ages of 12 and 17, since past research suggests this group uses the Internet the most and is victimized frequently (Mitchell et al., 2003; O'Connell et al., 2002; Sanger, et al., 2004; Wolak et al., 2002; Wolak et al., 2004; Wolak et al., 2006). The majority of the literature examined through this study also pertained to adolescents 12 to 17 years old. However, based on the difficulty associated with accessing this age group (because of the need for parental consent), this particular study utilized a sample of freshmen college students ages 18 and 19 years old. Although the age of 18 generally is accepted as the age of official adulthood, recent literature on brain development has asserted that sections of the brain associated with maturation and impulse control are not actually formed until well past the age of 18 (Giedd et al., 1999; Steinberg, 2004). This is an indication that the behavior of a 17 year old person is not necessarily much different than an 18 year old, as maturation change is not instantaneous, but instead constitutes a process.

Adolescence is a period of growth involving physiological and psychological development and maturation. During this time period, young people begin to take responsibility for their actions (Bandura, 1997; Bandura, 2000) and gain a sense of personal control of their lives (Catsambis, 1994; Wigfield, Eccles, Schiefele, Roeser, & Davis-Kean, 2006). This is attributed to changes in the frontal cortex of the brain, which is the area that is accountable for cognitive abilities, such as making plans, remembering tasks, and controlling inappropriate behavior (Huttenlocher, 1979). This change and development of the frontal cortex continues well beyond the childhood years and often into adulthood (Giedd et al., 1999), indicating maturation is a lengthy process that involves years of development, rather than an instantaneous transformation.

During this same phase of growing independence, there is a higher likelihood of risk-taking and impulsive behavior (Arnett, 1992). Adolescents tend to believe risks are smaller and more controllable, as compared to the perception of adults (Benthin, Slovis, & Severson, 1993). Adolescents also tend to participate in behaviors that may seem more risky at an older age, such as providing personal information to individuals online or engaging in offline contact with people met online. While many logical reasoning abilities appear to be developed by age 15, psychosocial attributes that enhance impulse control, decision-making, and resistance to peer influences continue to develop well into young adulthood (Steinberg, 2004). Again, this indicates that the full transition into adulthood is an extended process involving changes to the mental capacity past the age of 20.

The brief review of relevant literature presented above shows that the maturation process of adolescents is a progression spanning over the course of several years. Based on the noted empirical findings, it can be assumed the maturity of a 17-year-old will be similar to an 18 or 19-year-old. Therefore, the use of the latter age group for this study appears acceptable when examining Internet use and victimization, and when comparing the findings of the current research to past studies of adolescent online behavior.

During the fall 2007 semester, 2,673 freshmen students were enrolled at this particular university, which constituted 22.8% of the entire undergraduate population of 11,724. Since the 2003-2004 academic year, the number of freshmen enrolled at the university each academic year has declined slightly, so it was expected that the number of freshmen available for surveying in the spring 2008 semester was approximately 2,500.

The following is a list of all 100-level courses (according to the 2007-2008 University Undergraduate Catalog) potentially available to freshmen at the main campus in spring 2008, along with the respective sections available for each course. Students enrolled in these courses, which did not require a prerequisite, were eligible to participate in the survey:

Business and Technology Education
    Microbased Computer Literacy (BTED 101), Sections 001--002
    Microbased Computer Literacy (COSC 101), Sections 001–011
    Microbased Computer Literacy (IFMG 101), Sections 001–003

English
    College Writing (ENG 101), Sections 001 through 041

Fine Arts
    Introduction to Art (ARHI 101), Sections 001 through 006
    Introduction to Dance (DANC 102), Section 001
    Introduction to Music (MUHI 101), Sections 001 through 004
    Introduction to Theater (THTR 101), Sections 001 through 004

Health and Wellness
    Food and Nutrition (FDNT 143), Sections 001 through 002
    Healthy People (NURS 143), Sections 001 through 003
    Health and Wellness (HPED 143), Sections 001 through 015

History
    History: The Modern Era (HIST 195), Sections 001 through 029

Mathematics
    Intermediate Algebra (MATH 100), Sections 001 through 007
    Foundations of Mathematics (MATH 101), Sections 001– 012

Natural Science
    General Biology II and Lab (BIOL 104), Sections A01 through A06, B01 through B06, C01 through C06, D01 through D04, E01 through E06, F01 through 04
    College Chemistry I and Lab (CHEM 101), Sections A01 through A09
    College Chemistry II and Lab (CHEM 102), Sections B02, B03, B05, B07-B13

Philosophy and Religious Studies
    Introduction to Philosophy (PHIL 120), Sections 001 through 006
    Introduction to Religion (RLST 100), Sections 001 through 004
    World Religions (RLST 110), Sections 002, 005, and 006

Social Sciences
    American Politics (PLSC 111), Sections 001 through 006

Basic Economics (ECON 101), Section 001

Contemporary Anthropology (ANTH 110), Sections 001–010

Crime and Justice Systems (CRIM 101), Sections 001–003

Cultural Anthropology (ANTH 211), Sections 001–003

General Psychology (PSYC 101), Sections 001 through 015

Geography of Non-Western World (GEOG 104), Sections 001 through 011, and 013

Geography of the U.S. and Canada (GEOG 102), Sections 001 through 004

Introduction to Geography: Human Environment Interaction (GEOG 101), Sections 001 through 002

Principles of Sociology (SOC 151), Sections 001 through 013

World Politics (PLSC 101), Sections 001 through 009

This list of courses and the respective sections were placed in a table and numbered from 1 to 282 (the total sections available for selection in spring 2008). From this list, 75 sections were randomly selected via the use of a random number table. The professor of the particular section chosen was contacted via email with the following message:

Dear Professor _____,

My name is Catherine Marcum, and I am a doctoral candidate in the Department of Criminology. Your class, (class name), section (number), meeting on (days) from (time), has been randomly selected to participate in a student survey regarding freshmen Internet use and victimization. I would like to request your permission to administer the survey to your class at your convenience. It is expected that completion of the survey will take approximately 20 minutes.

Please let me know at your earliest convenience if you will allow the survey administration to commence. Thank you in advance for your time and assistance.

Sincerely,

Catherine Marcum

The random course selection continued until enough survey administrations were scheduled to produce the desired sample size. The following courses and corresponding number of sections were selected and scheduled for survey administration:

- American Politics (PLSC 111) – one section;
- Contemporary Anthropology (ANTH 110) – one section;
- Crime and Justice Systems (CRIM 101) – one section;
- General Psychology (PSYC 101) – six sections;
- Health and Wellness (HPED 143) – two sections;
- Healthy People (NURS 143) – three sections;
- History: The Modern Era (HIST 195) – one section;
- Introduction to Religion (RLST 100) – one section;
- Microbased Computer Literacy (COSC 101) – one section;
- World Politics (PLSC 101) – two sections.

From the survey administrations, a total of 850 surveys were obtained; however, only 483 were from freshmen and were used in the analysis for this study.

A basis for the desired sample size was established using Cohen's (1988) statistical power analysis equation for multiple regression techniques. The equation used to determine power is:

$$N = \frac{\lambda}{f^2}$$

where N represents the number of cases needed, λ (lambda) is the value of the noncentrality parameter of the noncentral F distribution, and f^2 is the effect size. The appropriate lambda value can be determined by referring to the lambda tables provided by Cohen; however, in order to do so, the number of independent variables (u), the degrees of freedom for error variance (v), and the desired power must be known.

The models for this study were planned to assess approximately 35 independent variables based on Routine Activities Theory, along with several control variables (age, gender, race, social bonds, parent-child conflict, and location of residence). The table does not specifically offer lambdas for 35 independent variables; therefore, a value of 40 independent variables (u) was used in formulating the equation. The degrees of freedom for error variance (v) are the power entries for each value for each independent variable (u). Based on the lambda tables, the choices for degrees of freedom for error variance are 20, 60, 120 and infinity. Cohen (1988) recommends the use of 120 as a trial value that will produce an N of sufficient size. Finally, in regard to a desired power value (i.e., the likelihood that a researcher will not commit a

Type II error by chance), Cohen suggests the use of 0.80 for behavioral science research. This will reduce the chance of making a Type II error to 0.20. Furthermore, when a model has not been analyzed previously, Cohen also recommends using by default a medium effect size (Cohen, 1988 p. 413). According to Cohen, it is reasonable for the value of 0.15 to define a medium effect size when there are numerous independent variables involved.

Using this information, the lambda (λ) value of 33.8 was obtained from Cohen's multiple regression sample size table with an alpha of .05. This resulted in the following equation:

$$N = \frac{33.8}{0.15} = 225.3$$

This indicates that a minimum sample size of 225 is needed in order to reliably uncover significant results. However, based on the components of this study, a larger sample was desirable than recommended by Cohen's (1988) equation. Estimations of the appropriate ratio of participants to independent variables also have been presented to assist in the determination of adequate sample size for regression analysis. Stevens (1992) asserted a ratio of 15 to 1 would be appropriate for a reliable regression equation, while Meyers, Gamst, and Guarino (2005) and Hair, Anderson, Tatham, and Black (1998) indicated a 20 to 1 ratio would be appropriate. Based on the number of possible independent variables in this study, these suggestions would encourage a sample of 500 to 700 subjects.

Other considerations also would suggest a larger sample size for this study. Generally, increasing the sample size allows researchers to produce lower standard errors and narrower confidence intervals (Meyers et al., 2005). A larger sample also will compensate for issues of non-response, missing values, and surveys discarded or not employed because of their unsuitability (i.e., completed by persons older than age 19 or above freshmen standing). The anticipated use of logistic regression in this study further necessitates a larger sample in order to produce reliable results through the statistical model (Hardy & Bryman, 2004). Finally, split models comparing males and females were utilized, requiring sufficient sample sizes within the two groups. Because of these issues and the calculations presented above, the sample of 483 obtained through the survey administration was considered appropriately large enough for analysis.

SURVEY CONSTRUCTION

In order to test the hypotheses, a survey was constructed to obtain information about respondents' use of the Internet and their victimization as a result of certain behaviors during Internet use, as well as their experiences with the formation of relationships with online contacts. Freshmen responded to survey items regarding these behaviors and experiences within two periods of time in their lives: as a high school senior and as a college freshman. Through the administration of the survey, the three central elements of Routine Activities Theory were measured: exposure to motivated offenders, target suitability, and lack of capable guardianship. The final survey utilized is presented in Appendix A. Codings for the variables are presented below in Tables 1 through 5.

Independent Variables

The first element of Routine Activities Theory evaluated was exposure to motivated offenders, which occurred through the examination of independent variables representing general usage of the Internet and specific modes of computer-mediated communication. First, students were asked the following questions regarding their general Internet use as high school seniors, and later as college students at a mid-sized university in the northeast:

1) How many hours a day did/do you typically spend on the Internet?

2) How many days a week did/do you use the Internet?

Next, questions regarding the types of activities performed online were accompanied by a set of pre-selected responses. Students first were asked to mark which, if any, of the following Internet activities they performed as a high school senior, and now as a college student at the university: research, gaming, planning travel, website design, shopping, socializing with others, and other.

The following survey questions also addressed this concept by asking respondents to reveal their experiences with computer-mediated communication as a high school senior and college student at the university:

1) Did/do you use email? If you answered yes, how many hours per week did/do you spend using email?

2) Did/do you use instant messaging? If you answered yes, how many hours per week did/do you spend using instant messaging?

3) Did/do you use chat rooms? If you answered yes, how many hours per week did/do you spend using chat rooms?

4) Did/do you use social networking websites? If you answered yes, how many hours per week did/do you spend using social networking websites?

Respondents also were questioned on the type of social networking website used, if any, as a high school senior and college student at the university. In general, if a particular site is inhabited by more motivated offenders compared to another site, the respondent may increase his or her chance of victimization by use of that site. Initial codings for these variables are listed in Table 1 below.

The second element of Routine Activities Theory evaluated was target suitability, which occurred through the examination of independent variables representing behaviors that indicate attractiveness as a suitable target for victimization. The following survey questions addressed this concept by asking respondents to reveal their behaviors as a high school senior and college student at the university:

1) Was/is your social networking website marked "private," so only designated friends could/can see your profile?

2) What types of information did/do you post on your social networking website?

3) Did/do you communicate with people online, via email, instant messaging, or chat rooms, that you had/have never met in person?

Table 1. Independent Variables Representing Exposure to Motivated Offenders and Codings

Days a week on the Internet	Use of instant messaging
0 – 7 days	No = 0
Activities performed on the Internet	Yes = 1
Research	Hours of use of instant messaging
No = 0	0 – 168 hours
Yes = 1	Use of chat rooms
Gaming	No = 0
No = 0	Yes = 1
Yes = 1	Hours of use of chat rooms
Planning travel	0 – 168 hours
No = 0	Use of social networking website
Yes = 1	No = 0
Website design	Yes = 1
No = 0	Hours of use of social networking
Yes = 1	website
Shopping	0 – 168 hours
No = 0	Social networking website used
Yes = 1	MySpace
Socializing with others	No = 0
No = 0	Yes = 1
Yes = 1	Facebook
Other	No = 0
No = 0	Yes = 1
Yes = 1	Other
Use of email	No = 0
No = 0	Yes = 1
Yes = 1	
Hours of use of email	
0 – 168 hours	

4) Did/do you voluntarily give personal information to a person you met online? If yes, please mark all types of information you gave to a person you met online via email, instant messaging, or chat rooms.

Codings for these variables are listed in Table 2 below.

Table 2. Independent Variables Representing Target Suitability and Coding

Privatization of social networking
website
    No = 0
    Yes = 1
Information posted on social
networking website
    Age
        No = 0
        Yes = 1
    Gender
        No = 0
        Yes = 1
    Descriptive characteristics
        No = 0
        Yes = 1
    Picture(s) of yourself
        No = 0
        Yes = 1
    Telephone number
        No = 0
        Yes = 1
    Calendar/schedule
        No = 0
        Yes = 1
    Social security number
        No = 0
        Yes = 1
    Birth date
        No = 0
        Yes = 1
    Home address
        No = 0
        Yes = 1
    School location
        No = 0
        Yes = 1
    Extracurricular activities
        No = 0
        Yes = 1

    Family member information
        No = 0
        Yes = 1
    Goals/aspirations
        No = 0
        Yes = 1
    Sexual information
        No = 0
        Yes = 1
    Emotional/mental distresses
    and problem
        No = 0
        Yes = 1
    Family conflicts
        No = 0
        Yes = 1
    Other
        No = 0
        Yes = 1
Communicate with strangers
online
    No = 0
    Yes = 1
Personal information to others
    No = 0
    Yes = 1
Information given to person(s)
online
    Age
        No = 0
        Yes = 1
    Gender
        No = 0
        Yes = 1
    Descriptive characteristics
        No = 0
        Yes = 1

Table 2, cont.

Picture(s) of yourself	Family member information
No = 0	No = 0
Yes = 1	Yes = 1
Telephone number	Goals/aspirations
No = 0	No = 0
Yes = 1	Yes = 1
Calendar/schedule	Sexual information (e.g.,
No = 0	fantasies, desires,
Yes = 1	experiences)
Social security number	No = 0
No = 0	Yes = 1
Yes = 1	Emotional/mental distresses
Birth date	and problem
No = 0	No = 0
Yes = 1	Yes = 1
Home address	Family conflicts
No = 0	No = 0
Yes = 1	Yes = 1
School location	Other
No = 0	No = 0
Yes = 1	Yes = 1
Extracurricular activities	
No = 0	
Yes = 1	

The final element of Routine Activities Theory assessed was lack of capable guardianship. The independent variables represent the amount of monitoring experienced by respondents as high school seniors and college freshmen at the university, as well their experiences with protective measures while using the Internet. The following survey questions, accompanied by a set of pre-selected responses, addressed this concept by asking respondents to reveal their experiences as a high school senior and college student:

1) Where did/do you most often use a computer?

2) Please mark any of the parties listed that were/are typically in the same room with you when you used/use a computer?

3) Please mark all of the restrictions you had/have from your parent/guardian while using the Internet.

4) To your knowledge, did/do your parent/guardian or another adult actively monitor your Internet use by regularly checking the websites you visited?

5) To your knowledge, was/is any type of blocking or filtering software on the computer(s) you typically used/use to protect you from unwanted materials?

Codings for these variables are listed in Table 3 below.

Table 3. Independent Variables Representing Lack of Capable Guardianship and Coding

Location of computer use
  Home
    No = 0
    Yes = 1
  What room?
    Living room/family room
      No = 0
      Yes = 1
    Your bedroom
      No = 0
      Yes = 1
    Parent/guardian's bedroom
      No = 0
      Yes = 1
    Other
      No = 0
      Yes = 1
  School computer lab
    No = 0
    Yes = 1
  Friend's home
    No = 0
    Yes = 1
  Coffee shop
    No = 0
    Yes = 1

  Other
    No = 0
    Yes = 1
In same room
  Parent/Guardian
    No = 0
    Yes = 1
  Friend
    No = 0
    Yes = 1
  Teacher/Counselor
    No = 0
    Yes = 1
  Sibling
    No = 0
    Yes = 1
  Someone else
    No = 0
    Yes = 1
  No one
    No = 0
    Yes = 1
Restrictions online
  Time spent online
    No = 0
    Yes = 1

Table 3, cont.

Viewing of adult websites	Active monitoring
No = 0	No = 0
Yes = 1	Yes = 1
Use of email, instant	Unsure if actively monitored
messaging, chat rooms, and	No = 0
social networking websites	Yes = 1
No = 0	No filtering/blocking software
Yes = 1	No = 0
Other	Yes = 1
No = 0	Filtering/blocking software used
Yes = 1	No = 0
No restrictions	Yes = 1
No = 0	Unsure if filtering/blocking
Yes = 1	software used
No active monitoring	N = 0
No = 0	Y = 1
Yes = 1	

Dependent Variables

For this study, there were two categories of dependent variables assessed. First, the concept of online victimization was defined as incidents occurring to persons that involve unwanted non-sexual harassment, unwanted exposure to sexual material, and solicitation for sexual activity.   Answers to survey questions initially were used to form dichotomous dependent variables. The frequency of these incidents also was requested, so the impact of the independent variables potentially could be examined on the frequency of the particular form(s) of victimization. The following survey questions addressed these variables with regard to respondent experiences as a high school senior and college student at the university:

1) Did/do you receive unwanted sexually explicit material over the Internet (e.g., pornographic pictures of naked people or people having sex)? On average, how many times per week did/does this occur?

2) Did/do you send sexually explicit material (e.g., porno-graphic pictures of naked people or people having sex) to

people over the Internet? On average, how many times per week did/does this occur?

3) Were/are you harassed in a non-sexual manner on the Internet (e.g., repetitive, unwanted emails or instant messages)? On average, how many times per week did/does this occur?

4) Did/do you harass people in a non-sexual manner on the Internet (e.g., repetitive, unwanted emails or instant messages)? On average, how many times per week did/does this occur?

5) Were/are you asked for sex on the Internet? On average, how many times per week did/does this occur?

6) Did/do you ask people for sex on the Internet? On average, how many times per week did/does this occur?

Second, additional items were used to measure dependent variables involving the formation of relationships with online contacts. The following survey questions addressed these experiences as a high school senior and college student at the university:

7) Did/do you form virtual relationships (i.e., a friendship or romantic relationship in which communication only occurred on the Internet) with people you met/meet online?

8) Did/do you participate in offline contact (i.e., communication not on the Internet) with people you met/meet on the Internet? A follow-up question requested the types of offline communication participated in with the online contact.

9) Did the offline relationship ever turn into a sexual encounter? A follow-up question asked if it was a willing sexual activity.

Codings utilized for these variables are listed in Table 4 below.

Table 4. Dependent Variables and Coding

Received unwanted sexually
explicit material
    No = 0
    Yes = 1
Frequency of receipt of unwanted
sexually explicit material
    0 = 0 times
    1 = 1-2 times per week
    2 = 3 or more times per week
Received non-sexual harassment
    No = 0
    Yes = 1
Frequency of receipt of non-sexual
harassment
    0 = 0 times
    1 = 1-2 times per week
    2 = 3 or more times per week
Received solicitation for sex
    No = 0
    Yes = 1
Frequency of receipt of
solicitation for sex
    0 = 0 times
    1 = 1-2 times per week
    2 = 3 or more times per week
Formed virtual relationship
    No = 0
    Yes = 1
Participated in offline contact
    No = 0
    Yes = 1

Forms of offline communication
Mail via U.S. Postal Service
    = 0
    Yes = 1
Called on the telephone
    No = 0
    Yes = 1
Received money or gifts from
person
    No = 0
    Yes = 1
Met at person's home
    No = 0
    Yes = 1
Met at your home
    No = 0
    Yes = 1
Met at other location
    No = 0
    Yes = 1
Other
    No = 0
    Yes = 1
Sexual encounter
    No = 0
    Yes = 1
Willing sexual activity
    No = 0
    Yes = 1

The purpose of introducing control variables to a study is to remove the effects of these particular variables so that the extent to which the independent variables affect the dependent variable can be accurately assessed (Punch, 2003). Control variables used in this study included sex, age, race, location of residence, social bonding variables, and

parent-child conflict variables. Codings for these variables are listed in
Table 5 below.

Table 5. Control Variables and Coding

Sex
  Male = 0
  Female = 1
Age
  Under 19 years of age = 0
  19 years of age or older = 1
Race
  White Non-Hispanic
    Not White Non-Hispanic = 0
    White Non-Hispanic = 1
  American Indian or Alaska
  Native
    Not American Indian or
    Alaska Native = 0
    American Indian or Alaska
    Native = 1
  African-American
    Not African-American = 0
    African-American = 1
  Asian
    Not Asian = 0
    Asian = 1
  Hispanic
    Not White Hispanic = 0
    White Hispanic = 1
  Other
    Not Other = 0
    Other = 1
Current Living Situation
  Living with parent(s)/guardian(s)
    Does not live with
    parent(s)/guardian(s) = 0
    Does live with
    parent(s)/guardian(s) = 1

Living with other family
member(s) besides parent(s)
  Does not live with other family
  member(s) besides parent(s) = 0
  Does live with other family
  member(s) besides parent(s) = 1
Living in a dormitory
  Does not live in a dormitory = 0
  Does live in a dormitory = 1
Living with friends in a rented
apartment/house
  Does not live with friends in a
  apartment/house = 0
  Does live with friends in a
rented apartment/house = 1
Living in a Greek
fraternity/sorority house
  Does not live in a Greek
  fraternity/sorority house = 0
  Does live in a Greek
  fraternity/sorority house = 1
Other
  Does not live in other location
  = 0
  Does live in other location = 1
High School GPA
  All A's = 7
  Mostly A's and B's = 6
  Mostly B's = 5
  Mostly B's and C's = 4
  Mostly C's = 3
  Mostly C's and D's = 2
  Mostly D's = 1
  Mostly D's and F's = 0

Table 5, cont.

Social Bonding
  I could/can share my thoughts and feelings with my parents/guardians.
    0.0 – 10.0
  I could/can share my thoughts and feelings with my friends.
    0.0 – 10.0
  I enjoyed/enjoy spending time with my friends.
    0.0 – 10.0
  I had/have respect for my parents/guardians.
    0.0 – 10.0
  I had/have respect for my teachers.
    0.0 – 10.0
  Participation in school activities was/is important to me.
    0.0 – 10.0

  I tried/try to stay involved in activities at school.
    0.0 – 10.0
  Getting good grades was/is important to me.
    0.0 – 10.0
  I tried/try hard to succeed at school.
    0.0 – 10.0
Parent-child Conflict
  My parents/guardians often yelled/yell at me.
    0.0 – 10.0
  My parents/guardians often nagged/nag me.
    0.0 – 10.0
  My parents/guardians often took/take away my privileges.
    0.0 – 10.0

Respondents' demographics (sex, age, and race) were controlled during the analysis. The sample employed was restricted to university freshmen who were 18 or 19 years of age, so age was controlled as a dichotomous variable. In regard to location of residence, Osgood et al. (1996) found that the absence of authority and protective figures in an adolescent's life increases the likelihood of deviant behavior. Students enrolled in college have various living situations, many of which do not involve the presence of parents or guardian figures. Effects of this variable were controlled while assessing the impact of guardianship variables previously presented in the survey.

With regard to the social bonding variables, Bernburg and Thorlindsson (2001) argued that routine activities are rooted in social contexts and relations, which are highly related to deviant behavior. In other words, social bonds formed through relationships with family and peers affect the patterning of activities and criminal behavior, and youth with stronger social bonds are less likely to participate in delinquent activities (Stewart, 2003). In general, social bonding items in the survey questioned respondents on their attachment, commitment, involvement, and beliefs (Hirschi, 1969).

Students first were questioned on their grade point average at the time of high school graduation. Secondly, through the use of visual analog scales (DeVellis, 2003), respondents were asked about their social bonds with parents, peers, and the school environment as a high school senior and college student. A majority of the statements measuring social bonding were used similarly by Durkin, Wolfe, and Clark (1999) in their study of binge drinking by college students, or were derived from other empirical studies that found higher levels of social bonding to be associated with lower levels of deviant behavior (Hirschi, 1969; Leonard & Decker, 1994; Michaels & Meithe, 1989; Stewart, 2003). The answer choices on the visual analog scales ranged from "strongly disagree" to "strongly agree." These scales are useful, as they do not limit a respondent to pre-set choices (as does a categorical Likert scale), resulting in a better representation of the respondent's opinion (DeVellis, 2003). The statements representing social bonding variables included the following:

1) I could/can share my thoughts and feelings with my parents.

2) I could/can share my thoughts and feelings with my friends.

3) I enjoyed/enjoy spending time with my friends.

4) I had/have respect for my parents.

5) I had/have respect for my teachers.

6) Participation in school activities was/is important to me.

7) I tried/try to stay involved in activities at school.

8) Getting good grades was/is important to me.

9) I tried/try hard to succeed at school.

Also asked in the form of visual analog scales were the survey items that measured parent-child conflict while the respondent was a high school senior, and now as a college student. Past studies have indicated that children with high levels of conflict with parents or guardians are more likely to participate in online relationships with strangers, which potentially could lead to victimization (Mitchell, Finkelhor, & Becker-Blease, 2007; Wolak et al., 2002; Wolak et al.,

2003; Ybarra, Mitchell, Finkelhor, & Wolak, 2007). The following items were used in both administrations of the Youth Internet Safety Survey and were used in this study to measure parent-child conflict:

1) My parents/guardians often yelled/yell at me.
2) My parents/guardians often nagged/nag me.
3) My parents/guardians often took/take away my privileges.

HUMAN SUBJECTS ISSUES

Participation in this study involved several human subjects issues for survey respondents (Babbie, 2004; Bachman & Paternoster, 1997; Bachman & Shutt, 2001; Dillman, 2007), which were addressed throughout the administration of the project. Regarding age, participants were required to be at least 18 years old, so the issue of parental consent for participation was not applicable. Any freshmen in the surveyed classes who were not at least 18 years old were asked not to accept or fill out a survey.

Participation in the study was completely voluntary. Students were informed verbally by the survey administrator, as well as notified in the Informed Consent Form, that participation was voluntary and not necessary for completion of any class requirement. Respondents were informed that if they initially chose to participate, and then decided against completing the survey at any time during the survey administration, they could write "withdraw" at the top of the survey and it would be destroyed.

Students were notified that no identifying information was being requested from them on the survey, nor should they make any notations that would provide evidence of their identity. These measures were to ensure anonymity of the participants. All completed surveys were kept secure in a locked filing cabinet in the researcher's home, so that only the researcher was able to access them.

If a student chose to participate in the study and complete the survey, it was believed that the potential risks were minimal. However, because of the sensitive nature of the topic (involving personal victimization), if a participant felt any type of physical or emotional stress caused by completing the survey, a sheet was attached to the back of each survey with contact information for mental health providers. It is attached as Appendix C.

If students had any questions before or during survey administration, they were encouraged to ask the survey administrator. No information was withheld from the participants regarding the purpose of the study and the contents of the survey.

ANALYSIS PLAN

The first step in the analysis plan was to compute univariate statistics and bivariate correlations. Frequencies and descriptive statistics were produced on all variables to examine the characteristics of the sample and the shape of the distribution for each variable (Bachman & Paternoster, 1997). Bivariate correlations then were calculated for all variables to assess the statistical associations among them. The most widely used bivariate correlation statistic is the Pearson correlation, often referred to as Pearson's r (Meyers, Gamst, & Guarino, 2006), which was used in this study. Pearson's r measures the extent of covariation between two variables. When independent variables were shown to be highly correlated, suggesting a strong linear relationship between them, they were considered for combination into one variable and implemented into subsequent models accordingly. A generally accepted standard for high correlation is a Pearson's r of 0.7 to 0.9 (Hardy & Bryman, 2004).

Since the dependent variables initially were measured as a dichotomy, logistic regression models were used to assess the relationship between the independent variables and the likelihood of victimization and formation of relationships with online contacts. Unlike other forms of regression analysis, logistic regression allows for the examination of the relationship between a categorical dependent variable and independent variables that are dichotomous or continuous (Menard, 2001; Meyers et al., 2006). Logistic regression also provides coefficients that allow for estimation of the effect that various independent variables have on the dependent variable, while controlling for other variables and testing for statistical significance (Hardy & Bryman, 2004).

Model estimation in logistic regression requires the probability of an event to be transformed into odds. The following equation can be utilized to estimate the probability of the dependent variable occurring:

$$\hat{P}(Y=1) = e^{a + b_1 x_1 + b_2 x_2 \ldots\ldots b_k x_k} / 1 - e^{a + b_1 x_1 + b_2 x_2 \ldots\ldots b_k x_k}$$

where "P" equals the estimated probability of being victimized or participating in a relationship with an online contact. The "a" is representative of the constant, "x" and "b" represent the independent variables and their corresponding slopes, and "e" equals the base of the natural logarithm (Hardy & Bryman, 2004; Menard, 2001; Meyers et al., 2006). In order to estimate a logistic regression equation, the logit function is applied:

$$Ln(P(Y=1)/1-P(Y=1)) = a + b_1x_1 + b_2x_2 + \ldots\ldots\ldots b_kx_k$$

In this equation, "a" is the predicted log odds of the dependent variable occurring when the independent variables are equal to zero, and "bx" represents the increase or decrease in the log odds of the dependent variable that is associated with specific values of the independent variables and their slopes. Comparatively, logistic regression is linear regression that uses logit as the outcome (Hardy & Bryman, 2004; Menard, 2001; Meyers et al., 2006).

Interpreting the log odds of an event can be difficult, so the natural log can be removed from logistic regression coefficients (B) to transform them into simple odds [Exp (B)]. The odds ratios (or exponentiated coefficients) show changes in the simple odds of the dependent variable occurring based on a one-unit increase in each independent variable. Positive coefficients, with odds ratios greater than 1.0, indicate that an increase in the independent variable is associated with a greater likelihood of the dependent variable occurring, while negative coefficients, with odds ratios less than 1.0, demonstrate an increase in the independent variable is linked with a lower likelihood of the dependent variable occurring (Hardy & Bryman, 2004; Menard, 2001; Meyers et al., 2006). In general, the models will show the change in the simple odds of online victimization and formation of relationships with online contacts occurring based on a one-unit increase in each independent variable, while controlling for the others.

The models developed for this study were based on the theoretical constructs of Routine Activities Theory, and they were utilized to examine the effects of various independent variables on the likelihood of online victimization and formation of relationships with online contacts for respondents both as high school seniors and college students at the university. For each reference to online victimization, models were built to examine the effect of the independent variables on each type of dependent variable. Models also were built to examine the effect of independent variables on the respondents' formation of online and .

First, the effects of the independent variables measuring the theoretical construct of exposure to motivated offenders were assessed. This was done through the use of logistic regression models and initially addressed Hypothesis 1. To illustrate, concerning the use of the four modes of computer-mediated communication (CMCs), the models indicated whether greater use of each of the CMCs increased the likelihood of online victimization and relationship formation.

The next set of models assessed the addition of the effects of the independent variables measuring the theoretical construct of target suitability, while also including variables measuring exposure to motivated offenders. These models addressed Hypotheses 2 thorough 4. For example, a logistic regression model considered if providing certain types of identifying personal information (shown in Table 2) to online contacts increased the likelihood of online victimization and relationship formation. The effect of providing this personal information to an online contact was assessed, along with the effect of leaving a social networking website available for public viewing.

Full models eventually were constructed to assess the addition of lack of capable guardianship variables, while including measures of exposure to motivated offenders and target suitability, and subsequently controlling for gender, age, race, location of residence social bonding variables, and parent-child conflict variables. This addressed Hypotheses 5 through 7 and provided more complete tests of Hypotheses 1 through 4. These models examined such things as the effect of utilizing filtering and blocking software on online victimization, the effect of location of the computer used most frequently, and the effect of various persons monitoring the Internet use of the respondent. For example, the effect of using a computer in the family room versus other locations (where it can be assumed there will be less guardianship) was examined with regard to the likelihood of victimization and relationship formation. The effect of different types of parent/ guardian restrictions and monitoring also was assessed (e.g., restricting time spent online and restricting viewing of adult websites).

As previously mentioned, separate models also were estimated for males and females to consider the differences, if any, between the sexes regarding their predictors of online victimization and the formation of relationships with online contacts. The logistic regression models noted above were utilized separately for males and females, and the results from each of the models allowed for an assessment of whether certain factors have greater or lesser impact on online victimization and relationship formation for males as compared to females.

SUMMARY

The purpose of this study was to examine Internet use and types of use by freshmen at a mid-sized university in the northeast, in relation to various behaviors and experiences. Since there is currently a gap in the literature with regard to research that tests relationships between adolescents' online use and their resulting experiences, this study sought to provide an important contribution regarding this emerging topic area.

The chosen methodology was developed under the concepts and propositions of Routine Activities Theory, which has been utilized many times in the past to explain various forms of victimization. The use of this empirically supported theory to examine the behaviors and experiences of college freshmen on the Internet was expected to clarify which variables increase or decrease the likelihood of victimization and formation of relationships with online contacts by the respondents.

An original survey was constructed for use in this study, to collect information about Internet use and the experiences of respondents as high school seniors and college freshmen. The data collected through the survey were analyzed through the use of univariate statistics, bivariate correlations, and logistic regression models. Independent variables representing exposure to motivated offenders, target suitability, and lack of guardianship were inserted into the models to determine their effect, if any, on online victimization and formation of relationships with online contacts. Multiple models were utilized to determine if there is a higher or lower likelihood of online victimization and formation of relationships with online contacts as a result of certain behaviors online, and if so, the degree of impact of these variables.

It was hypothesized that the findings would reveal spending higher amounts of time on the Internet in certain locations and providing personal information to online contacts increases the likelihood of victimization and formation of relationships with online contacts for adolescents, while various measures of guardianship decrease the likelihood of these outcomes. From the study's findings and contributions, it was anticipated that preventative policies and educational programs could be suggested to reduce the occurrence of online victimization and better inform adolescents and parents about the formation of relationships with online contacts.

CHAPTER 5:

Univariate and Bivariate Results

In an effort to produce a more complete understanding of adolescent Internet use, online victimization, and formation of relationships with online contacts, 483 completed surveys were collected from freshmen enrolled at a university in the northeast. The surveys generated data from respondents on their frequency and types of Internet use, development of relationships with online contacts, and experiences with different types of Internet victimization at two periods in their lives: as a high school senior and as a college freshman. This chapter outlines the univariate and bivariate results from the analysis of this data. First, a brief discussion is provided of the pilot study performed before the actual survey administration. Second, frequency and descriptive data produced from the survey administration are examined. Finally, bivariate correlations among the various independent and dependent variables are considered.[1]

PILOT STUDY

Before actual administration of the survey to the freshmen classes, a pilot study was performed to test the reliability and validity of the survey instrument. The choice of a pilot study was made for two reasons: 1) based on the lack of explanatory studies on the topic, there

---

[1] It should be noted that when bivariate correlations and subsequent multivariate models were estimated, the use of mean imputation was utilized to resolve the issue of missing data. Past studies (Little, 1992; Sande, 1982) support the use of mean imputation as a method of substituting for missing data, especially when the amount of missing data is small. In this study, there was an average of 1% missing data across the measures, and no more than 3% for any item.

was little past research to rely upon for reliability and validity purposes; and 2) the survey instrument used in this study was an original instrument; therefore, it was not previously tested for reliability and validity. With the use of a pilot study, the survey instrument was evaluated and amended before actual administration to the sample of freshmen.

The pilot study included administration of the entire survey to the four sections of undergraduate research methods classes offered to criminology undergraduates in the fall 2007 semester. Completion of the survey took approximately 20 minutes. From this administration, 83 completed surveys were obtained, with the majority of the pilot study sample (72%) being juniors. Fifty-nine of the respondents (69.5%) were male.

After the surveys were collected, participants were given an opportunity to ask questions and provide suggestions for improvement of the survey. While several different suggestions for change were made, those ideas proposed most often were used to amend the survey instrument. In general, changes were made to make the survey more concise and for clarification purposes. Revisions were made to some of the questions pertaining to victimization, so that the respondent would know exactly what was considered to be victimization in the survey. For example, questions asking about experiences with unwanted exposure to sexual materials were clarified with the addition of the phrase "not including pop-ups," so the respondent would be clear on what sexual materials were referenced in the question. Additions also were made to the second part of the survey to more accurately reflect college experiences. For instance, the choice of "dorm room" was added to the list of places a respondent could select as a location where he or she most often uses the computer.

The surveys then were coded and entered for statistical analysis. Basic analyses were conducted to check for measurement problems and assess if items were performing as expected. For example, an examination of the frequency distributions for the target suitability variables revealed that several types of information that could be posted on a social networking website or given to contacts online were not chosen by any of the respondents. Several of the pilot study respondents also stated that the length of the list of posted information was too long.

Therefore, choices that were not selected by any survey participants (e.g., SSN and home address) were removed from the survey.

Bivariate correlations then were calculated to further assess the need, if any, to combine or eliminate variables representing the same theoretical construct. Survey items measuring the number of hours per day on the Internet and number of hours per week on the Internet (measures of exposure to motivated offenders) were highly correlated for experiences in both the time periods of high school senior ($r = .808$, $p < .01$) and college freshman ($r = .869$, $p < .01$). The survey item measuring number of hours per week was therefore removed from the survey. No other measures were found to be as highly correlated, so no other items were eliminated.

Finally, preliminary models planned for analysis of the freshmen sample were run on the pilot study data. Overall, the data performed as expected, indicating that certain independent variables significantly influenced the likelihood of different types of online victimization and relationship formation for pilot study respondents. While using a relatively small pilot study sample, however, some of the independent variables had no significant effect. It was expected that with a larger sample of freshmen, low statistical power would be less of an issue within the analysis.

FREQUENCIES AND DESCRIPTIVES

The frequencies and descriptives were examined first for the variables pertaining to experiences as a high school senior, and then as a college freshman. Each section examining the two separate time periods was also categorized based on the theoretical constructs and dependent variables. Finally, the frequency distributions of the control variables (i.e., demographic characteristics of the sample) were considered, along with the descriptive measures of the social bonding and parent-child conflict variables.

Responses Pertaining to Experiences as a High School Senior

Exposure to motivated offenders. Frequencies for the categorical variables measuring exposure to motivated offenders are shown in Table 6. Descriptive statistics for the continuous variables are presented in Tables 7 and 8.

Table 6. Frequencies for categorical variables representing exposure to motivated offenders while a high school senior (N = 483)

Variable	N	%	Variable	N	%
Activities performed on the Internet			Use of email (n = 482)		
			No	91	18.9
Research (n = 482)			Yes	391	81.1
No	23	4.8	Use of instant messaging (n = 482)		
Yes	459	95.2	No	93	19.3
Gaming (n = 482)			Yes	389	80.7
No	223	46.3	Use of chat rooms (n = 482)		
Yes	259	53.7	No	442	91.7
Planning travel (n = 482)			Yes	40	8.3
No	326	67.6	Use of social networking		
Yes	156	32.4	websites (n = 482)		
Website design (n = 482)			No	89	18.5
No	406	84.2	Yes	393	81.5
Yes	76	15.8	Social networking website used		
Shopping (n = 482)			MySpace (n = 480)		
No	193	40.0	No	178	37.1
Yes	289	60.0	Yes	302	62.9
Socializing with others (n = 482)			Facebook (n = 480)		
No	47	9.8	No	180	37.5
Yes	435	90.2	Yes	300	62.5
Other (n = 481)			Other (n = 480)		
No	429	89.2	No	464	96.7
Yes	52	10.8	Yes	16	3.3

Table 7. Descriptive Statistics for Continuous Variables Representing Exposure to Motivated Offenders While a High School Senior (N = 483)

Variable	Minimum	Maximum	Mean	SD
Hours per day on the Internet (n = 479)	0	20	2.85	2.33
Days per week on the Internet (n = 481)	0	7	5.92	1.54
Hours per week of use of email (n = 479)	0	48	1.87	3.21
Hours per week of use of instant messaging (n = 480)	0	122	6.62	9.40
Hours per week of use of chat rooms (n = 480)	0	12	0.22	1.06
Hours per week of use of social networking websites (n = 482)	0	70	5.54	7.18

Table 8. Descriptive Statistics for Recoded Continuous Variables Representing Exposure to Motivated Offenders While a High School Senior (N = 483)

Variable	Minimum	Maximum	Mean	SD
Hours per week on the Internet (n = 479)	0	35	15.14	8.97
Hours per week of use of email (n = 479)	0	4	1.29	1.06
Hours per week of use of instant messaging (n = 480)	0	15	4.39	4.09
Hours per week of use of chat rooms (n = 480)	0	1	0.08	0.27
Hours per week of use of social networking websites (n = 477)	0	15	4.05	3.79

As shown in Tables 6 and 7, the respondents reported performing various activities while using the Internet that potentially exposed them to motivated offenders. Respondents reported spending an average of 2.85 hours per day and 5.92 days per week online when they were high school seniors. Four hundred fifty-nine (95.2%) performed research, 259 (53.7%) used it for gaming, 156 (32.4%) planned travel, 76 (15.8%) designed websites, 289 (60%) shopped, 435 (90.2%) socialized with others, and 52 (10.8%) reported using the Internet for other purposes.

Survey respondents also reported on their use of four types of computer-mediated communications: email, instant messaging, chat rooms, and social networking websites. While a high school senior, 391 (81.1%) subjects used email, with a mean of 1.87 hours per week; 389 (80.7%) used instant messaging, with a mean of 6.62 hours per week; 40 (8.3%) used chat rooms, with a mean of 0.22 hours per week; and 393 (81.5%) used social networking websites, with a mean of 5.54 hours per week. Regarding the specific social networking websites used, 302 (62.9%) used MySpace, 300 (62.5%) used Facebook, and 16 (3.3%) used another social networking website.

Table 8 shows the continuous variables created from the recoding of values in order to produce a more normal distribution for these items. All of these variables were created due to a severe positive skew in the original variables. Through an examination of the initial distributions, all extreme values in the tails were collapsed.[2]

First, the measure of hours per week on the Internet was created by multiplying the values of hours per day on the Internet and days per week on the Internet. Since the other continuous variables were measured in the time span of a week, this new variable also seemed better suited for implementation in the models. Values past 35 hours per week were recoded to equate to 35 hours. Based on this recoded variable, respondents reported spending an average of 15.14 hours per week online when they were high school seniors. For the variable hours per week of use of email, any value past 4 hours was collapsed and recoded to equate to 4 hours. Based on this recoded variable, respondents reported spending an average of 1.29 hours using email. For instant messaging use, any value past 15 hours was collapsed to equate to 15 hours. Based on this

---

[2] Skew and kurtosis statistics were used to assess the normality of the distributions.

recoded variable, respondents reported spending an average of 4.39 hours using instant messaging. For hours per week of chat room use, values past 1 hour per week were recoded to equate to 1 hour. Based on the recoded variable, respondents reported spending an average of 0.08 hours per week using chat rooms. Finally, for social networking website use, any value past 15 hours was collapsed to equate to 15 hours for social networking website use. Based on the recoded variable, respondents reported spending an average of 4.05 hours per week using social networking websites.

Target suitability

Frequencies for the categorical variables measuring target suitability are presented in Table 9. Table 9 examines behaviors and information provided by high school seniors that potentially makes them suitable targets for victimization and relationship formation while online. There were 237 respondents (49.3%) who used a non-privatized social networking website. The other 244 (50.7%) either did privatize their website or did not use a social networking website. The frequencies and corresponding types of personal information provided on social networking websites included: 361 (75.1%) posted their age, 390 (81.1%) posted their gender, 126 (26.2%) posted descriptive characteristics, 383 (79.6%) posted a picture of themselves, 29 (6.0%) posted a telephone number, 260 (54.1%) posted their school location, 290 (60.3%) posted their extracurricular activities, 144 (29.9%) posted goals/aspirations, 10 (2.1%) posted sexual information, 30 (6.2%) posted emotional and mental distresses, 7 (1.5%) posted family conflicts, and 26 (5.4%) posted other types of information.

Two hundred and seven (43.2%) respondents reported communicating with strangers online, and 100 (20.7%) reported giving personal information to others online while a high school senior. Types of information provided to others online were as follows: 101 (21%) disclosed their age, 104 (21.6%) disclosed their gender, 57 (11.8%) disclosed descriptive characteristics, 60 (12.4%) provided pictures of themselves, 38 (7.9%) gave their telephone number, 43 (8.9%) disclosed their school location, 66 (13.7%) disclosed extracurricular activities, 43 (8.9%) disclosed goals/aspirations, 15 (3.1%) disclosed sexual information, emotional and mental distresses, or family conflicts, and 2 (0.4%) disclosed other types of information.

Table 9. Frequencies for Categorical Variables Representing Target Suitability While a High School Senior (N = 483)

Variable	N	%
Used a non-privatized social networking website (n = 481)		
No	244	50.7
Yes	237	49.3
Information posted on social networking website		
Age (n = 481)		
No	120	24.9
Yes	361	75.1
Gender (n = 481)		
No	91	18.9
Yes	390	81.1
Descriptive characteristics (n = 481)		
No	355	73.8
Yes	126	26.2
Picture(s) of yourself (n = 481)		
No	98	20.4
Yes	383	79.6
Telephone number (n = 481)		
No	452	94.0
Yes	29	6.0
School location (n = 481)		
No	221	45.9
Yes	260	54.1
Extracurricular activities (n = 481)		
No	191	39.7
Yes	290	60.3
Goals/aspirations (n = 481)		
No	337	70.1
Yes	144	29.9
Sexual information (n = 481)		
No	471	97.9
Yes	10	2.1
Emotional/mental distresses/ problems (n = 481)		
No	451	93.8
Yes	30	6.2

Family conflicts (n = 481)		
Variable	**N**	**%**
No	474	98.5
Yes	7	1.5
Other (n = 481)		
No	455	94.6
Yes	26	5.4
Communicate with strangers online (n = 479)		
No	272	56.8
Yes	207	43.2
Personal information to others (n = 482)		
No	382	79.3
Yes	100	20.7
Information given to person(s) online		
Age (n = 482)		
No	381	79.0
Yes	101	21.0
Gender (n = 482)		
No	378	78.4
Yes	104	21.6
Descriptive characteristics (n = 482)		
No	425	88.2
Yes	57	11.8
Picture(s) of yourself (n = 482)		
No	422	87.6
Yes	60	12.4
Telephone number (n = 482)		
No	444	92.1
Yes	38	7.9
School location (n = 482)		
No	439	91.1
Yes	43	8.9
Extracurricular activities (n = 482)		
No	416	86.3
Yes	66	13.7

Table 9, continued

Variable	N	%	Variable	N	%
Goals/aspirations (n = 482)			Family conflicts (n = 482)		
No	439	91.1	No	467	96.9
Yes	43	8.9	Yes	15	3.1
Sexual information (n = 483)			Other (n = 482)		
No	468	97.1	No	480	99.6
Yes	15	3.1	Yes	2	0.4
Emotional/mental distresses/					
problems (n = 482)					
No	467	96.9			
Yes	15	3.1			

Lack of capable guardianship

Frequencies for the categorical variables measuring guardianship are shown in Table 10. As shown in Table 10, there were certain behaviors and experiences of respondents as high school seniors that indicate the theoretical construct of lack of capable guardianship. First, the location of computer use was examined. Four hundred and forty-seven (92.9%) respondents stated they used the computer a majority of the time in the home of a parent or guardian, in several different locations. One hundred ninety-four (40.8%) used in the living/family room, 155 (32.6%) used in their own bedroom, 8 (1.7%) used in a parent/guardian's bedroom, and 81 (17.1%) used in another room in the home. As for the rest of the survey respondents, 22 (4.6%) used in a school computer lab, 6 (1.3%) used at a friend's home, and 8 (1.7%) used in another location.

Survey respondents then were asked to name the people typically in the room with them when they used a computer. Two hundred and twenty-five (46.8%) were accompanied by a parent/guardian, 258 (53.6%) were accompanied by a friend, 66 (13.7%) were accompanied by a teacher/counselor, 223 (46.4%) were accompanied by a sibling, 52 (10.8%) were accompanied by someone else, and 206 (42.8%) had no one else in the room with them.

Table 10. Frequencies for Categorical Variables Representing Lack of Capable Guardianship While a High School Senior (N = 483)

Variable	N	%	Variable	N	%
Location of computer use			Sibling (n = 481)		
Home (n = 481)			No	258	53.6
No	34	7.1	Yes	223	46.4
Yes	447	92.9	Someone else (n = 481)		
Living room/family room			No	429	89.2
(n = 475)			Yes	52	10.8
No	281	59.2	No one (n = 481)		
Yes	194	40.8	No	275	57.2
Your bedroom (n = 475)			Yes	206	42.8
No	320	67.4	Restrictions online		
Yes	155	32.6	Time spent online (n = 480)		
Parent/guardian's bedroom			No	404	84.2
(n = 475)			Yes	76	15.8
No	467	98.3	Viewing of adult websites		
Yes	8	1.7	(n = 480)		
Other room (n = 475)			No	309	64.4
No	394	82.9	Yes	171	35.6
Yes	81	17.1	Use of CMCs (n = 480)		
School computer lab (n = 480)			No	453	94.4
No	458	95.4	Yes	27	5.6
Yes	22	4.6	Other (n = 480)		
Friend's home (n = 480)			No	467	97.3
No	474	98.8	Yes	13	2.7
Yes	6	1.3	No restrictions (n = 480)		
Coffee shop (n = 480)			No	216	45.0
No	480	100.0	Yes	264	55.0
Yes	0	0.0	No active monitoring (n = 478)		
Other (n = 480)			No	184	38.5
No	472	98.3	Yes	294	61.5
Yes	8	1.7	Active monitoring (n = 478)		
In same room			No	411	86.0
Parent/Guardian (n = 481)			Yes	67	14.0
No	256	53.2	Unsure of active monitoring		
Yes	225	46.8	(n = 478)		
Friend (n = 481)			No	361	75.5
No	223	46.4	Yes	117	24.5
Yes	258	53.6	No filtering/blocking software		
Teacher/Counselor (n = 481)			(n = 478)		
No	415	86.3	No	291	60.9
Yes	66	13.7	Yes	187	39.1

Table 10, cont.

Variable	N	%	Variable	N	%
Filtering/blocking software (n = 478)			Unsure of filtering/blocking software (n = 478)		
No	239	50.0	No	426	89.1
Yes	239	50.0	Yes	52	10.9

A portion of the survey respondents noted restrictions given to them by a parent or guardian while using the Internet. Seventy-six (15.8%) had time restrictions, 171 (35.6%) had viewing restrictions of adult websites, 27 (5.6%) had use of computer-mediated communications restrictions, 13 (2.7%) had other restrictions, and 264 (55%) had no restrictions. Two hundred ninety-four (61.5%) respondents reported they were not actively monitored during Internet use, 67 (14.0%) reported they were actively monitored during Internet use, and 117 (24.5%) reported they did not know if they were actively monitored during Internet use. With regard to the use of filtering and blocking software, 187 (39.1%) reported they did not use the software, 239 (50%) did use the software, and 52 (10.9%) were unsure if they used the software.

Dependent variables

Frequencies for the categorical dependent variables are shown in Table 11. Table 11 demonstrates that a somewhat sizeable portion of the survey respondents experienced one or more types of online victimization while a high school senior. One hundred and eight respondents (22.8%) reported receiving unwanted sexually explicit material. Sixty respondents (12.7%) received it a reported 1-2 times per week, and 48 (10.1%) received it 3 or more times per week. One hundred forty-four (30.8%) reported receiving non-sexual harassment. Fifty-seven (12.2%) received it a reported 1-2 times per week, and 87 (18.6%) received it 3 or more times per week. Finally, 45 respondents (9.6%) reported receiving sexual solicitation. Thirty-five respondents (7.4%) received it an average of 1-2 times per week, and 10 (2.1%) received it an average of 3 or more times per week.

Table 11. Frequencies for Categorical Variables Representing Dependent Variables While a High School Senior (N = 483)

Variable	N	%	Variable	N	%
Received unwanted sexually explicit material (n = 473)			Sent non-sexual harassment (n = 481)		
No	365	77.2	No	469	97.5
Yes	108	22.8	Yes	12	2.5
Frequency of receiving unwanted sexually explicit materials (n = 473)			Sent solicitation for sex (n = 481)		
			No	473	98.3
0 = No receipt	365	77.2	Yes	8	1.7
1 = 1-2 times per week	60	12.7	Formed virtual relationship (n = 481)		
2 = 3 or more times per week	48	10.1	No	387	80.5
			Yes	94	19.5
Received harassment in non-sexual manner (n = 468)			Participated in offline contact (n = 481)		
No	324	69.2	No	396	82.3
Yes	144	30.8	Yes	85	17.7
Frequency of receiving harassment in non-sexual manner (n = 468)			Forms of offline contact		
			Mail via U.S. Postal Service (n = 481)		
0 = No receipt	324	69.2	No	469	97.5
1 = 1-2 times per week	57	12.2	Yes	12	2.5
2 = 3 or more times per week	87	18.6	Called on the telephone (n = 481)		
Received solicitation for sex (n = 470)			No	413	85.9
			Yes	68	14.1
No	425	90.4	Money or gifts (n = 481)		
Yes	45	9.6	No	474	98.5
Frequency of receiving solicitation for sex (n = 470)			Yes	7	1.5
			Met at person's home (n = 481)		
0 = No receipt	425	90.4	No	461	95.8
1 = 1-2 times per week	35	7.4	Yes	20	4.2
2 = 3 or more times per week	10	2.1	Met at your home (n = 481)		
			No	468	97.3
Sent sexually explicit material (n = 481)			Yes	13	2.7
			Met at other location (n = 481)		
No	465	96.7	No	430	89.4
Yes	16	3.3	Yes	51	10.6
			Other (n = 481)		
			No	470	97.7
			Yes	11	2.3

Table 11, cont.

Variable	N	%	Variable	N	%
Sexual encounter (n = 481)			Willing sexual activity (n = 481)		
No	461	95.8	No	461	95.8
Yes	20	4.2	Yes	20	4.2

Also shown in Table 11 is the small number of survey respondents who reported one or more types of online offending while a high school senior. Sixteen respondents (3.3%) reported sending sexually explicit material, 12 (2.5%) reported sending non-sexual harassment, and 8 (1.7%) reported sending sexual solicitation. Based on the small amount of these types of behaviors reported, these variables were excluded from further bivariate and multivariate analysis.

Some of the respondents also became involved in different types of relationships with people they met online. Ninety-four (19.5%) formed a virtual relationship with a person online, and 85 (17.7%) participated in offline contact with a person they met online. With regard to offline contact, the following experiences were reported by respondents: 12 (2.5%) sent or received something via U.S. mail, 68 (14.1%) called or were called on the telephone, 7 (1.5%) sent or received money or gifts, 20 (4.2%) met at the person's home, 13 (2.7%) met a their own home, 51 (10.6%) met at another location, and 11 (2.3%) used another form of offline communication. Twenty (4.2%) students reported participating in a sexual encounter with an online contact, all of which were willing.

Responses Pertaining to Experiences as a College Freshman

Exposure to motivated offenders

Frequencies for the categorical variables measuring exposure to motivated offenders are shown in Table 12, and descriptive statistics for the continuous variables are presented in Tables 13 and 14.

As shown in Tables 12 and 13, the respondents reported performing various activities while using the Internet. As a college freshman, respondents reported spending an average of 4.21 hours per day and 6.49 days per week online. Four hundred seventy-three (98.1%) performed research, 223 (46.3%) used it for gaming, 151 (31.3%) planned travel, 64 (13.3%) designed websites, 299 (62%) shopped, 444 (92.1%) socialized with others, and 71 (14.7%) reported using the Internet for other purposes.

Survey respondents also reported on their use of four types of computer-mediated communications: email, instant messaging, chat rooms, and social networking websites. As a college freshman, 479 (99.4%) subjects used email, with a mean of 3.16 hours per week; 391 (81.1%) used instant messaging, with a mean of 7.38 hours per week; 23 (4.8%) used chat rooms, with a mean of 0.13 hours per week; and 458 (94.8%) used social networking websites, with a mean of 7.83 hours per week. With regard to the specific social networking websites used, 274 (56.8%) used MySpace, 430 (89.2%) used Facebook, and 16 (3.3%) used another social networking website.

Table 12. Frequencies for Categorical Variables Representing Exposure to Motivated Offenders While a College Freshman (N = 483)

Variable	N	%	Variable	N	%
Activities performed on the Internet			Use of email (n = 482)		
			No	3	0.6
Research (n = 482)			Yes	479	99.4
No	9	1.9	Use of instant messaging (n = 482)		
Yes	473	98.1			
Gaming (n = 482)			No	91	18.9
No	259	53.7	Yes	391	81.1
Yes	223	46.3	Use of chat rooms (n = 482)		
Planning travel (n = 482)			No	459	95.2
No	331	68.7	Yes	23	4.8
Yes	151	31.3	Use of social networking websites (n = 483)		
Website design (n = 482)					
No	418	86.7	No	25	5.2
Yes	64	13.3	Yes	458	94.8
Shopping (n = 482)			Social networking website used		
No	183	38.0	MySpace (n = 482)		
Yes	299	62.0	No	208	43.2
Socializing with others (n = 482)			Yes	274	56.8
			Facebook (n = 482)		
No	38	7.9	No	52	10.8
Yes	444	92.1	Yes	430	89.2
Other (n = 482)			Other (n = 482)		
No	411	95.3	No	466	96.7
Yes	71	14.7	Yes	16	3.3

**Table 13. Descriptive Statistics for Continuous Variables
Representing Exposure to Motivated Offenders While a
College Freshman (N = 483)**

Variable	Minimum	Maximum	Mean	SD
Hours per day on the Internet (n = 479)	0	20	4.21	3.44
Days per week on the Internet (n = 475)	1	7	6.49	1.05
Hours per week of use of email (n = 475)	0	21	3.16	3.24
Hours per week of use of instant messaging (n = 479)	0	148	7.38	13.94
Hours per week of use of chat rooms (n = 479)	0	10	0.13	0.83
Hours per week of use of social networking websites (n = 469)	0	148	7.83	11.79

**Table 14. Descriptive Statistics for Recoded Continuous Variables
Representing Exposure to Motivated Offenders While a
College Freshman (N = 483)**

Variable	Minimum	Maximum	Mean	SD
Hours per week on the Internet (n = 479)	0	50	26.64	14.19
Hours per week of use of email (n = 475)	0	8	2.54	1.58
Hours per week of use of instant messaging (n = 475)	0	14	4.86	4.95
Hours per week of use of chat rooms (n = 479)	0	1	0.06	0.20
Hours per week of use of social networking websites (n = 469)	0	15	6.05	4.90

Table 14 shows the continuous variables created from the recoding of values in order to produce a more normal distribution for these items. All of these variables were created due to a severe positive skew in the original variables. Through an examination of the initial distributions, all extreme values in the tails were collapsed.[3]

First, the measure of hours per week on the Internet was created by multiplying the values of hours per day on the Internet and days per week on the Internet. Values past 50 hours per week were recoded to equate to 50 hours. Based on the recoded variable, respondents reported spending an average of 26.64 hours per week online as college freshmen. For the variable hours per week of use of email, any value past 8 hours was recoded to equate to 8 hours. Based on this recoded variable, respondents reported spending an average of 2.54 hours using email. For instant messaging use, any value past 14 hours was collapsed to equate to 14 hours. Based on the recoded variable, respondents reported spending an average of 4.86 hours using instant messaging. For use of chat rooms, values past 1 hour per week were recoded to equate to 1 hour. Based on this recoded variable, respondents reported spending an average of 0.06 hours using chat rooms. Finally, any value past 15 hours was collapsed to equate to 15 hours for social networking website use. Based on the recoded variable, respondents reported spending an average of 6.05 hours using social networking websites.

Overall, in this study, college freshmen were shown to use the Internet more extensively as compared to high school seniors. Respondents also reported that they used each individual CMC more as a college freshman than as a high school senior. This examination of these variables suggests greater exposure to motivated offenders as college freshmen.

Target suitability

Frequencies for the categorical variables measuring target suitability are shown in Table 15.

Table 15 presents behaviors and information provided by college freshmen that potentially makes them suitable targets for victimization while online. There were 200 (41.5%) respondents who used a non-privatized social networking website. Types of personal information

---

[3] Skew and kurtosis statistics were used to assess normality of the distributions.

Table 15. Frequencies for Categorical Variables Representing Target Suitability While a College Freshmen (N = 483)

Variable	N	%
Uses a non-privatized social networking website (n = 482)		
No	282	41.5
Yes	200	58.5
Information posted on social networking website		
Age (n = 482)		
No	56	11.6
Yes	426	88.4
Gender (n = 482)		
No	31	6.4
Yes	451	93.6
Descriptive characteristics (n = 482)		
No	334	69.3
Yes	148	30.7
Picture(s) of yourself (n = 482)		
No	30	6.2
Yes	452	93.8
Telephone number (n = 482)		
No	408	84.6
Yes	74	15.4
School location (n = 482)		
No	92	19.1
Yes	390	80.9
Extracurricular activities (n = 482)		
No	180	37.3
Yes	302	62.7
Goals/aspirations (n = 482)		
No	300	62.2
Yes	182	37.8
Sexual information (n = 482)		
No	473	98.1
Yes	9	1.9
Emotional/mental distresses/problems (n = 482)		
No	456	94.6
Yes	26	5.4

Variable	N	%
Family conflicts (n = 482)		
No	476	98.8
Yes	6	1.2
Other (n = 482)		
No	444	92.1
Yes	38	7.9
Communicate with strangers online (n = 483)		
No	262	54.2
Yes	221	45.8
Personal information to others (n = 483)		
No	405	83.9
Yes	78	16.1
Information given to person(s) online		
Age (n = 483)		
No	405	83.9
Yes	78	16.1
Gender (n = 483)		
No	407	84.3
Yes	76	15.7
Descriptive characteristics (n = 483)		
No	436	90.3
Yes	47	9.7
Picture(s) of yourself (n = 483)		
No	435	90.1
Yes	48	9.9
Telephone number (n = 483)		
No	461	95.4
Yes	22	4.6
School location (n = 483)		
No	429	88.8
Yes	54	11.2
Extracurricular activities (n = 483)		
No	436	90.3
Yes	47	9.7

Table 15, cont.

Variable	N	%
Goals/aspirations (n = 483)		
No	448	92.8
Yes	35	7.2
Sexual information (n = 483)		
No	476	98.6
Yes	7	1.4
Emotional/mental distresses and problems (n = 483)		
No	473	97.9
Yes	10	2.1

Variable	N	%
Family conflicts (n = 483)		
No	475	98.3
Yes	8	1.7
Other (n = 483)		
No	478	99.0
Yes	5	1.0

provided on social networking websites were as follows: 426 (88.4%) posted their age, 451 (93.6%) posted their gender, 148 (30.7%) posted descriptive characteristics, 452 (93.8%) posted a picture of themselves, 74 (15.4%) posted a telephone number, 390 (80.9%) posted their school location, 302 (62.7%) posted extracurricular activities, 182 (37.8%) posted goals/aspirations, 9 (1.9%) posted sexual information, 26 (5.4%) posted emotional and mental distresses, 6 (1.2%) posted family conflicts, and 38 (7.9%) posted other types of information.

Two hundred twenty-one (45.8%) respondents reported communicating with strangers online, and 78 (16.1%) reported giving personal information to others online as a college freshman. Types of information provided to others online were as follows: 78 (16.1%) disclosed their age, 76 (15.7%) disclosed their gender, 47 (9.7%) disclosed descriptive characteristics, 48 (9.9%) gave pictures of themselves, 22 (4.6%) gave their telephone number, 54 (11.2%) disclosed their school location, 47 (9.7%) disclosed extracurricular activities, 35 (7.2%) disclosed goals/aspirations, 7 (1.4%) disclosed sexual information, 10 (2.1%) disclosed emotional and mental distresses, 8 (1.7%) disclosed family conflicts, and 5 (1.0%) disclosed other types of information.

After examination of the variables representing target suitability, there were various differences revealed between respondents as high school seniors and as college freshmen. As college freshmen, respondents were more likely to use social networking websites, but less likely to privatize the website. They also reported higher frequencies of providing personal information on their social networking websites.

High school seniors, however, reported higher frequencies of providing personal information to online contacts as compared to college freshmen.

Lack of capable guardianship

Frequencies for the categorical variables measuring guardianship are presented in Table 16. As shown in Table 16, there are certain behaviors and experiences of respondents as college freshmen that indicate the theoretical construct of lack of capable guardianship. It was expected that there would be differences, as compared to the high school senior time period, since many subjects would be living away from home with fewer restrictions and less guardianship over online use. First, the location of computer use was examined. Fifty-eight (12%) respondents stated they used the computer a majority of the time in the home of a parent or guardian, in several different locations. Twelve (2.5%) used in the living/family room, 37 (7.7%) used in his or her own bedroom, and 2 (0.4%) used in another room in the home. As for the rest of the survey respondents, 422 (87.4%) used in a dorm room, 7 (1.4%) used in a school computer lab, and 3 (0.6%) used at a friend's home.

Survey respondents then were asked to name the people typically in the room with them when they used a computer while a college freshman. Twenty (4.1%) were accompanied by a parent/guardian, 373 (77.2%) were accompanied by a friend, 10 (2.1%) were accompanied by a teacher/counselor, 26 (5.4%) were accompanied by a sibling, 85 (17.6%) were accompanied by someone else, and 177 (36.6%) had no one else in the room with them.

A portion of the survey respondents noted restrictions given to them by a parent or guardian while using the Internet. Seven (1.5%) had time restrictions, 45 (9.4%) had viewing restrictions of adult websites, 3 (0.6%) had use of computer-mediated communications restrictions, 17 (3.5%) had other restrictions, and 414 (85.9%) had no restrictions. Three hundred sixty-seven (76.1%) respondents reported they were not actively monitored during Internet use, 13 (2.7%) reported they were actively monitored during Internet use, and 102 (21.2%) reported they did not know if they were actively monitored during Internet use. With regard to the use of filtering and blocking software, 224 (46.6%) reported they did not use the software, 191 (39.7%) did use the software, and 66 (13.7%) were unsure if they used the software.

Table 16. Frequencies for Categorical Variables Representing Lack of Capable Guardianship While a College Freshman (N = 483)

Variable	N	%	Variable	N	%
Location of computer use			Teacher/Counselor (n = 483)		
Home (n = 483)			No	473	97.9
No	425	88.0	Yes	10	2.1
Yes	58	12.0	Sibling (n = 483)		
Living room/family room (n = 481)			No	457	94.6
No	469	97.5	Yes	26	5.4
Yes	12	2.5	Someone else (n = 483)		
Your bedroom (n = 481)			No	398	82.4
No	444	92.3	Yes	85	17.6
Yes	37	7.7	No one (n = 483)		
Parent/guardian's bedroom (n = 481)			No	306	63.4
No	481	100.0	Yes	177	36.6
Yes	0	0.0	Restrictions online		
Other (n = 481)			Time spent online (n = 481)		
No	479	99.6	No	474	98.5
Yes	2	0.4	Yes	7	1.5
Dorm (n = 483)			Viewing of adult websites (n = 481)		
No	61	12.6	No	436	90.6
Yes	422	87.4	Yes	45	9.4
School computer lab (n = 483)			Use of CMCs (n = 481)		
No	476	98.6	No	478	99.4
Yes	7	1.4	Yes	3	0.6
Friend's home (n = 483)			Other (n = 481)		
No	480	99.4	No	464	96.5
Yes	3	0.6	Yes	17	3.5
Coffee shop (n = 483)			No restrictions (n = 481)		
No	483	100.0	No	67	13.9
Yes	0	0.0	Yes	414	85.9
Other (n = 483)			No active monitoring (n = 482)		
No	483	100.0	No	115	23.9
Yes	0	0.0	Yes	367	76.1
In same room			Active monitoring (n = 482)		
Parent/Guardian (n = 483)			No	469	97.3
No	463	95.9	Yes	13	2.7
Yes	20	4.1	Unsure of active monitoring (n = 482)		
Friend (n = 483)			No	380	78.8
No	110	22.8	Yes	102	21.2
Yes	373	77.2			

Table 16, cont.

Variable	N	%	Variable	N	%
No filtering/blocking software (n = 481)			Unsure of filtering/blocking software (n = 481)		
No	257	53.4	No	415	86.3
Yes	224	46.6	Yes	66	13.7
Filtering/blocking software (n = 481)					
No	290	60.3			
Yes	191	39.7			

Dependent variables

Frequencies for the categorical dependent variables are shown in Table 17. Table 17 demonstrates that, as compared to the high school senior time period, a smaller portion of the survey respondents experienced one or more types of online victimization while a college freshman. Forty-eight respondents (10.0%) reported receiving unwanted sexually explicit material. Twenty-seven of those respondents (5.6%) received it an average of 1-2 times per week, and 21 respondents (4.4%) received it an average of 3 or more times per week. Sixty-two (12.9%) reported receiving non-sexual harassment. Nineteen (4.0%) received it an average of 1-2 times per week, and 43 (9.0%) received it an average of 3 or more times per week. Finally, 28 respondents (5.8%) reported receiving sexual solicitation. Fourteen respondents (2.9%) received it an average of 1-2 times per week, and 14 (2.9%) received it an average of 3 or more times per week.

Also shown in Table 17 is the extremely small portion of survey respondents who reported one or more types of online offending while a college freshman. Six respondents (1.2%) reported sending sexually explicit materials, 4 (0.8%) reported sending non-sexual harassment, and 4 (0.8%) reported sending sexual solicitation. Based on the small amount of these behaviors reported, which was even smaller than that reported for the high school senior time period, these variables were excluded from further analysis.

Table 17. Frequencies for Categorical Dependent Variables While a College Freshman (N = 483)

Variable	N	%
Received unwanted sexually explicit material (n = 480)		
No	432	90.0
Yes	48	10.0
Frequency of receiving unwanted sexually explicit material (n = 480)		
0 = No receipt	432	90.0
1 = 1-2 times per week	27	5.6
2 = 3 plus times per week	21	4.4
Received harassment in a non-sexual manner (n = 480)		
No	418	87.1
Yes	62	12.9
Frequency of receiving non-sexual harassment (n = 480)		
0 = No receipt	418	87.1
1 = 1-2 times per week	19	4.0
2 = 3 plus times per week	43	9.0
Received solicitation for sex (n = 477)		
No	454	94.2
Yes	28	5.8
Frequency of receiving solicitation for sex (n = 480)		
0 = No receipt	454	95.2
1 = 1-2 times per week	14	2.9
2 = 3 plus times per week	14	2.9
Sent sexually explicit material (n = 483)		
No	477	98.8
Yes	6	1.2
Sent non-sexual harassment (n = 483)		
No	479	99.2
Yes	4	.8
Sent sexual solicitation (n = 483)		
No	479	99.2
Yes	4	.8
Formed virtual relationship (n = 482)		
No	390	80.9
Yes	92	19.1
Participated in offline contact (n = 482)		
No	403	83.6
Yes	79	16.4
Forms of offline contact		
Mail via U.S. Postal Service (n = 482)		
No	476	98.8
Yes	6	1.2

Table 17, cont.

Variable	N	%
Called on the telephone (n = 482)		
No	431	89.4
Yes	51	10.6
Received money or gifts from person (n = 482)		
No	477	99.0
Yes	5	1.0
Met at person's home (n = 482)		
No	464	96.3
Yes	18	3.7
Met at your home (n = 482)		
No	469	97.3
Yes	13	2.7
Met at other location (n = 482)		
No	435	90.2
Yes	47	9.8
Other (n = 482)		
No	467	96.9
Yes	15	3.1
Sexual encounter (n = 482)		
No	470	97.5
Yes	12	2.5
Willing sexual activity (n = 482)		
No	470	97.5
Yes	12	2.5

Some of the respondents also became involved in different types of relationships with people they met online. Ninety-two (19.1%) formed a virtual relationship with a person online, and 79 (16.4%) participated in offline contact with a person they met online. With regard to offline contact, the following experiences were reported by the respondents: 6 (1.2%) sent or received something via U.S. mail, 51 (10.6%) called or were called on the telephone, 5 (1.0%) sent or received money or gifts, 18 (3.7%) met at the person's home, 13 (2.7%) met a their own home, 47 (9.8%) met at another location, and 15 (3.1%) used another form of offline communication. Twelve (2.7%) people reported participating in a sexual encounter with an online contact, all of which were willing.

Control variables
Table 18 shows the frequencies for the control variables for the sample of freshmen. The sample was composed of 288 (60.1%) females and 191

males (39.9%). Two hundred forty-four (51.3%) respondents were 18 years old, and the rest (48.7%) were 19 years old. The race variable consisted of 401 (83.7%) white non-Hispanics, 3 (0.6%) American Indian or Alaskan natives, 36 (7.5%) African-Americans, 4 (0.8%) Asians, 19 (4.0%) white Hispanics, and 16 others (3.3%). In regard to living situation, 37 (7.7%) lived with a parent/guardian, 2 (0.4%) lived with other family members, 410 (85.6%) lived in a dormitory, 27 (5.6%) lived in a rented apartment/ house, and 3 (0.6%) lived in another location. Finally, grade point averages at the time of high school graduation varied.

Table 18. Frequencies for Control Variables (N = 483)

Variable	N	%
Sex (n = 479)		
Male	191	39.9
Female	288	60.1
Age (n = 476)		
18 years old	244	51.3
19 years old	232	48.7
Race (n = 479)		
White Non-Hispanic	401	83.7
American Indian or Alaska Native	3	0.6
African-American	36	7.5
Asian	4	0.8
White Hispanic	19	4.0
Other	16	3.3
Current Living Situation (n = 479)		
Living with parent(s)/guardian(s)	37	7.7
Living with other family member(s) besides parent(s)	2	0.4
Living in a dormitory	410	85.6
Living with friends in a rented apartment/house	27	5.6
Other	3	0.6
GPA (n = 476)		
All A's	43	9.0
Mostly A's and B's	233	48.9
Mostly B's	82	17.2
Mostly B's and C's	91	19.1
Mostly C's	20	4.2
Mostly C's and D's	4	0.8
Mostly D's	1	0.2
Mostly D's and F's	2	0.4

Forty-three (9%) respondents reported having all A's, 233 (48.9%) had mostly A's and B's, 82 (17.2%) had mostly B's, 91 (19.1%) had mostly B's and C's, 20 (4.2%) had mostly C's, 4 (0.8%) had mostly C's and D's, 1 (0.2%) had mostly D's, and 2 (0.4%) had mostly D's and F's.

After examination of the frequencies in Table 18, two variables were collapsed for implementation in subsequent models, due to the large majority of responses falling into one category. First, with regard to race, a large percentage of respondents (83.7%) reported they were white non-Hispanic. Two categories subsequently were created for analysis: White and Nonwhite. Second, with regard to living situation, 85.6% reported they currently lived in a dormitory, and the remainder of the respondents either lived with family or in another situation. Three categories subsequently were created for analysis: living in a dorm, living with family, and other living arrangements.

Table 19 portrays descriptive statistics for the items used to measure social bonding and parent-child conflict. The first set of 12 items asked respondents to rate how much they agreed or disagreed with statements pertaining to their time as a high school senior. Attachment was measured in the first three items, with a mean of 6.10 for "I could share my thoughts and feelings with parents," a mean of 7.70 for "I could share my thoughts and feelings with friends," and a mean of 8.46 for "I enjoyed spending time with friends." Belief was measured with the statements "I had respect for my parents," scoring a mean of 7.93, and "I had respect for my teachers," scoring a mean of 7.46. Next, involvement was measured with a mean of 6.61 on the statement "Participation in school activities was important to me," and a mean of 6.09 for the statement "I tried to stay involved in activities at school." Commitment was measured with the statements "Getting good grades was important to me," with a mean score of 7.51, and "I tried hard to succeed in school," with a mean score of 6.95. Finally, parent-child conflict variables were measured with the statements "My parents/guardians often yelled at me," with a mean of 3.88; "My parents/guardians often nagged me," with a mean of 4.82; and "My parents/guardians often took privileges away" with a mean of 2.84.

Table 19. Descriptive Statistics for Social Bonding and Parent/Child Conflict Variables (N = 483)

Variable	Minimum	Maximum	Mean	SD
High School				
I could share my thoughts and feelings with parents.				
(n = 473)	0.0	10.0	6.10	2.66
I could share my thoughts and feelings with friends.				
(n = 473)	0.0	10.0	7.70	1.87
I enjoyed spending time with my friends.				
(n = 473)	0.0	10.0	8.46	1.39
I had respect for my parents.				
(n = 472)	0.2	10.0	7.93	1.92
I had respect for my teachers.				
(n = 472)	0.6	10.0	7.46	1.96
Participation in school activities was important to me.				
(n = 472)	0.0	10.0	6.61	2.78
I tried to stay involved in activities at school.				
(n = 472)	0.0	10.0	6.09	2.85
Getting good grades was important to me.				
(n = 472)	0.0	10.0	7.51	2.11
I tried hard to succeed at school.				
(n = 472)	0.0	10.0	6.95	2.45
My parents/guardians often yelled at me.				
(n = 472)	0.0	10.0	3.88	2.81
My parents/guardians often nagged me.				
(n = 471)	0.0	10.0	4.82	2.93
My parents/guardians often took away privileges.				
(n = 471)	0.0	10.0	2.84	2.63
College				
I can share my thoughts and feelings with parents.				
(n = 471)	0.0	10.0	7.36	2.26
I can share my thoughts and feelings with friends.				
(n = 471)	0.0	10.0	8.06	1.79
I enjoy spending time with my friends.				
(n = 470)	0.0	10.0	8.50	1.45
I have respect for my parents.				
(n = 471)	0.0	10.0	8.16	1.74
I have respect for my teachers.				
(n = 470)	1.2	10.0	8.11	1.43
Participation in school activities is important to me.				
(n = 472)	0.0	10.0	5.87	2.65
I try to stay involved in activities at school.				
(n = 472)	0.0	10.0	5.12	2.77

Table 19, cont.

Variable	Minimum	Maximum	Mean	SD
Getting good grades is important to me.				
(n = 472)	0.0	10.0	8.51	1.37
I try hard to succeed at school.				
(n = 471)	0.2	10.0	8.29	1.52
My parents/guardians often yell at me.				
(n = 472)	0.0	10.0	2.53	2.42
My parents/guardians often nag me.				
(n = 472)	0.0	10.0	3.30	2.83
My parents/guardians often take away privileges.				
(n = 472)	0.0	10.0	1.34	1.75

The second set of 12 items asked respondents to rate how much they agreed or disagreed with statements pertaining to their time as a college freshman. Attachment was measured in the first three items, with a mean of 7.36 for "I can share my thoughts and feelings with parents," a mean of 8.06 for "I can share my thoughts and feelings with friends," and a mean of 8.50 for "I enjoy spending time with friends." Belief variables were measured with the statements "I have respect for my parents," scoring a mean of 8.16, and "I have respect for my teachers," scoring a mean of 8.11. Next, involvement was measured with a mean of 5.87 on the statement "Participation in school activities is important to me," and a mean of 5.12 for the statement "I try to stay involved in activities at school." Commitment was measured with the statements "Getting good grades is important to me," with a mean score of 8.51, and "I try hard to succeed in school," with a mean score of 8.29. Finally, parent-child conflict variables were measured with the statements "My parents/guardians often yell at me," with a mean of 2.53; "My parents/guardians often nag me" with a mean of 3.30; and "My parents/guardians often take privileges away" with a mean of 1.34.

BIVARIATE RESULTS

This section will present a summary of the estimated correlations among the variables used in the study. Although the values represent statistical associations between two variables without controlling for the potential effect of other variables, these correlations still can provide a preliminary description of the general relationships between the variables. As there are a large number of variables examined, and in turn a sizeable number of correlations, significant correlations with a moderate to strong relationship (±0.40 and above) will be emphasized in this section. The correlations for each time period measured in the study (i.e., high school senior and college freshman) are discussed separately for each set of variables.

Exposure to Motivated Offenders

Although there were a number of statistically significant correlations among the variables representing the theoretical construct of exposure to motivated offenders during the high school senior time period, several emerged that had a moderately strong correlation of ±0.40 or above. Overall, these correlations indicated positive associations between use of CMCs, time spent on the Internet, and activities performed while online.

As expected, there were strong correlations between the types of CMCs utilized and respective hours per week spent in each CMC, indicating that the reported use of a specific CMC was associated with increased time using the CMC. Use of email (Email) was strongly associated with hours per week spent on email (EmHours) ($r = .673$, p $< .01$); use of instant messaging (IM) was similarly associated with hours per week spent on instant messaging (IMHours) ($r = .555$, p $< .01$); use of chat rooms (Chat) was highly associated with hours per week spent in chat rooms (ChatHours) ($r = .970$, p $< .01$); and use of social networking websites (SNW) was positively associated with hours per week spent in social networking websites (SNWHours) ($r = .554$, p $< .01$). Also, hours per week spent on instant messaging (IMHours) had a moderate correlation with hours per week spent on social networking websites (SNWHours) ($r = .426$, p $< .01$), which indicated increasing use of social networking websites was associated with more time spent instant messaging p $< .01$), indicating greater Internet use was associated with greater instant messaging use. Hours per week use of social networking websites (SNWHours) had a similar correlation with hours per week use of email (EmHours) ($r = .435$, p $< .01$), as well as hours per week use of instant messaging (IMHours) ($r = .537$, p $< .01$), indicating that increased time using social networking websites was associated with spending more time using other CMCs. Also with regard to the use of social networking websites, the use of a social networking website (SNW) had a strong association with the use of Facebook (r = .643, p < .01), indicating that college freshmen were more likely to use Facebook over other types of social networking websites.

Finally, as expected, there were positive correlations between the reported use of CMCs and the respective hours per week of use of the CMC. Use of instant messaging (IM) had a moderately strong correlation with hours spent per week using instant messaging (IMHours) (r = .474, p < .01). There was a perfect correlation (r = 1.000, p < .01) between the use of chat rooms (Chat) and hours spent in a chat room per week (ChatHours), due to recoding of the latter variable as a dichotomy.

Target Suitability

This section discusses the correlations between independent variables representing the theoretical construct of target suitability during the high school senior time period. There again were many statistically significant correlations between these variables, with multiple associations emerging that had at least a correlation of ±0.40 or above. Overall, respondents who communicated with people online were likely to provide personal information online as well. There also were strong relationships between different types of information either posted on a social networking website or provided to online contacts, indicating that providing one type of information increased the likelihood of providing another type of information.

The correlation matrix displays substantial associations between several types of personal information posted on social networking websites, indicating that respondents were likely to provide more than one type of personal information on their social networking website. For example, respondents who posted their age on social networking websites (AgeSNW) also were very likely to post their gender (GenSNW) (r = .824, p < .01) and a picture (PicSNW) (r = .804, p < .01). Posting a picture (PicSNW) was strongly correlated with posting gender (GenSNW) (r = .941, p < .01). In a similar manner, providing age to online contacts (Age) was highly associated with providing gender (Gen) (r = .975, p < .01) and a picture (Pic) (r = .732, p < .01). Providing a picture (Pic) was also strongly correlated with providing gender (Gen) (r = .708, p < .01).

There were positive associations between various other types of information, continuing to demonstrate that providing one type of information increased the likelihood of providing other types of information. For example, posting age on a social networking website (AgeSNW) was correlated with posting school location (SchSNW) (r = .507, p < .01) and extracurricular activities (r = .494, p < .01). Also, providing age to online contacts (Age) was associated with providing a telephone number (Tel) (r = .568, p < .01) and goals (Goal) (r = .607, p < .01) to online contacts. Finally, providing a telephone number to online contacts (Tel) was associated with providing school location (Sch) (r = .448, p < .01) and emotional distresses (Emot) (r = .479, p < .01).

Other notable correlations also emerged from this matrix. Respondents who communicated with online contacts (Comm) were more likely to provide personal information to these online contacts (Personal) (r = .512, p < .01). Providing personal information to online contacts (Personal) was strongly associated with providing age (Age) (r = .975, p < .01) and gender (Gen) (r = .963, p < .01). In other words, respondents who reported they provided personal information to online contacts were very likely to have provided their age and gender. Interestingly, there was a negative correlation between using a non-privatized social networking site (NoPrivSNW) and posting gender (GenSNW) (r = -.470, p < .01) and a picture (PicSNW) (r = -.474, p < .01). In general, using a non-privatized social networking website reduced the likelihood a respondent posted these types of personal information.

The next correlations explored were between independent variables representing the theoretical construct of target suitability during the college freshman time period. As with the high school senior time period, there were correlations between several types of personal information, indicating that respondents who provided personal information were likely to provide more than one type on their social networking website or to online contacts. For example, posting age on a social networking website (AgeSNW) had fairly strong correlations with posting gender (GenSNW) (r = .618, p < .01) and a picture (PicSNW) (r = .630, p < .01). There were also strong associations between providing these same types of information to online contacts. Providing age to online contacts (Age) was highly correlated with providing gender (Gen) (r = .969, p < .01) and a picture (Pic) (r = .757, p < .01), again indicating that a person who provided age also was likely to provide gender and a picture to an online contact.

There were moderately strong associations among various other types of information, continuing to demonstrate that providing one type of information increased the likelihood of providing other types of information. Providing descriptive characteristics (Des) to online contacts was strongly associated with providing multiple other types of information, such as age (Age) (r = .748, p < .01), a picture (Pic) (r = .685, p < .01), and extracurricular activities (Extra) (r = .646, p < .01). There were strong correlations between providing sexual information (Sex) and emotional distresses (Emot) (r = .834, p < .01), along with family conflicts (Fam) (r = .799, p < .01). Providing emotional information to online contacts (Emot) was also highly correlated with providing family conflict information (Fam) (r = .893, p < .01).

Finally, similar to the high school senior time period, respondents who communicated with online contacts (Comm) were more likely to provide personal information to these online contacts (Personal) (r = .444, p < .01). Providing personal information to online contacts (Personal) was strongly associated with multiple types of personal information, such as age (Age) (r = .924, p < .01), gender (Gen) (r = .907, p < .01), school location (r = .757, p < .01) and extracurricular activities (r = .729, p < .01). Therefore, a respondent who reported providing personal information to online contacts as a college freshman had an increased likelihood of providing a number of pieces of information.

Lack of Capable Guardianship

There were several correlations among independent variables representing the theoretical construct lack of capable guardianship during the high school senior time period. The first set of notable correlations pertains to the main location of computer use. Use of a home computer (HomeComp) was negatively correlated with use of the school computer lab (SchLab) (r = -.794, p < .01), as well as other locations (OthPl) (r = -.408, p < .01), indicating that (as expected) using a home computer for a respondent's main Internet use was inversely associated with using it elsewhere. Similarly, use of a living room computer (LivRm) was negatively associated with use in the respondent's bedroom (YourBed) (r = -.578, p < .01).

There also was an association between having a parent in the room and having a sibling in the room while the respondent was using the Internet. Having a parent in the room (ParInRm) was positively correlated with having a sibling in the room (SibInRm) (r = .543, p < .01). Therefore, the presence of a parent in the room increased the likelihood a sibling would be present as well.

Moreover, there were also correlations between independent variables representing the theoretical construct lack of capable guardianship during the college freshman time period. Again, much like the high school senior time period, the main location of computer use was found to be associated with other locations, and these correlations were the main notable associations for the college freshman time period. The main use of a home computer (HomeComp) was positively correlated with use in the living room (LivRm) (r = .426, p < .01), as well as in the respondent's bedroom (YourBed) (r = .803, p < .01). Therefore, if respondents used a home computer for their main Internet use, it likely would be the living room computer or (even more likely) a computer in their bedroom. The use of a dorm room computer (Dorm) was negatively correlated with the use of a home computer (HomeComp) (r = -.870, p < .01), indicating that mainly using a dorm room computer for Internet use was inversely associated with using a computer in the home.

Dependent Variables

After examination of the correlations between the dependent variables during the high school senior time period, in general, participating in offline contact was associated with various forms of victimization. Those respondents who participated in offline contact also were likely to participate in more than one type of contact.

There were several variables positively associated with participating in offline contact with a person met online (PartOffline), indicating that participating in offline contact was associated with an increased likelihood of other experiences. For example, there was at least a moderate correlation with forming a virtual relationship with an online contact (VirtRelation) (r = .429, p < .01), indicating that a person who formed a virtual relationship with someone he or she met online was more likely to report an offline contact. With regard to relationships between participating in offline contact with a person met online (PartOffline) and specific types of offline contact, there were also several substantial correlations with telephone contact (Telephone) (r = .857, p < .01), meeting at the home of the online contact (ContactHome) (r = .428, p < .01), and meeting at other locations (OtherLocation) (r = .700, p < .01). These correlations indicate that a respondent participating in offline contact was likely to have spoken to the online contact on the telephone or met them in person. Finally, there was a considerable association between participating in offline contact (PartOffline) and participating in a sexual encounter with the online contact (SexEnc) (r = .443, p < .01). This indicates a significant likelihood that participating in offline contact would include a sexual relationship with an online contact. There were also positive correlations between the various types of offline contact. Participating in a telephone conversation with an online contact (Telephone) was associated with meeting at the home of the online contact (ContactHome) (r = .461, p < .01), meeting at the home of the respondent (YourHome) (r = .417, p < .01), and meeting at other locations (r = .584, p < .01). From these correlations, respondents who participated in a telephone contact with an online contact were also likely to meet them in person. There was also a strong association between meeting at the home of an online contact (ContactHome) and meeting at the home of the respondent (YourHome) (r = .672, p < .01), indicating that respondents who met an online contact at the contact's home were also likely to have met them at their own home.

Finally, there were several variables that exhibited at least a moderate correlation with participation in a sexual encounter with an online contact (SexEnc), suggesting that participation in these behaviors was associated with an increased likelihood of participating in a sexual encounter with the online contact. For example, the use of the telephone to speak with an online contact (Telephone) had a positive association with participating in a sexual encounter (SexEnc) (r = .415, p < .01). Moreover, as expected, participation in a sexual encounter was consistently correlated with meeting someone in person. There were positive associations between participation in a sexual encounter (SexEnc) and meeting an online contact at his or her home (ContactHome) (r = .439, p < .01), the respondent's home (YourHome) (r = .427, p < .01), and at other locations (OtherLocation) (r = .425, p < .01).

There were also correlations between the dependent variables during the college freshman period. First, much like the high school time period, there were several variables positively associated with participating in offline contact with a person met online (PartOffline). There were substantial correlations with telephone contact (Telephone) (r = .783, p < .01), meeting at the home of the online contact (ContactHome) (r = .445, p < .01), and meeting at other locations (OtherLocation) (r = .737, p < .01). Therefore, respondents participating in offline contact with a person they met online were likely to do so over the telephone and in person. There was an additional association with forming a virtual relationship with an online contact (VirtRelation) (r = .590, p < .01), indicating that a person who formed a virtual relationship with someone he or she met online was likely to have a relationship offline. Persons who formed a virtual relationship with an online contact were also more likely to participate in offline contact via the telephone (Telephone) (r = .475, p < .01) and through meeting at other locations (OtherLocation) (r = .478, p < .01).

There were also observed correlations between various types of offline contact. Exchanging mail via the United States Postal System (Mail) was strongly correlated with the exchange of money or gifts (Money/gifts) (r = .727, p < .01), suggesting that if a respondent and online contact exchanged money or gifts, it was through the mail. Participating in a telephone conversation with an online contact (Telephone) was associated with meeting at the home of the online contact (ContactHome) (r = .493, p < .01) and meeting at other locations (r = .560, p < .01). From these correlations, respondents who participated in telephone contact with an online contact were also likely to meet them in person. There was also a strong association between meeting at the home of an online contact (ContactHome) and meeting at the home of the respondent (YourHome) (r = .691, p < .01), indicating that respondents who met an online contact at the contact's home were also likely to have met them at their own home.

Unlike the high school time period, there was only one strong correlation associated with the development of a sexual relationship. Meeting at the home of an online contact (ContactHome) was associated with participation in a sexual encounter (SexEnc) (r = .507, p <

.01). While correlations with other variables were statistically significant, there were no others that were at least ± 0.40.

Control Variables

There were few correlations between the control variables. The only notable statistically significant associations between the demographic variables concern the variables representing living situation. There was a strong negative correlation between living with family (LivFam) and between living in a dorm room (LivDorm) (r = -.743, p < .01), as well as living in a dorm room (LivDorm) and living in another location (LivOther) (r = -.615, p < .01). These associations would be expected, as a respondent living in one of the three locations would not be living in the other two locations.

The remainder of the table provides the correlations among the social bonding and parent-child conflict variables during the high school senior and college freshman time periods. As expected, there were several moderately strong associations between the variables within the same time period. For example, during the high school time period, there was a positive association between sharing thoughts and feelings with friends (HSShareFriend) and enjoying spending time with friends (HSEnjoyFriend) (r = .608, p < .01). Therefore, respondents who felt more comfortable sharing thoughts and feelings with friends were more likely to enjoy spending time with friends. There was also a strong positive association between the importance of getting good grades (HSGrades) and succeeding in school (HSSucceed) (r = .779, p < .01). This correlation indicated that respondents connected academic success with getting good grades while a high school senior. With regard to the college freshman time period, respect held for parents (ColRespParent) was associated with the ability to share thoughts and feelings with parents (ColShareParent) (r = .599, p < .01).

There were also significant correlations between variables across the two time periods. For example, involvement in high school activities (HSInvolved) was positively associated with involvement in college activities (ColInvolved) (r = .454, p < .01). There was also a strong positive correlation between the ability to share thoughts and feelings with parents during the high school senior time period (HSShareParent) and the ability to share thoughts and feelings with parents during the college freshman time period (ColShareParent) (r = .678, p < .01).

A few extremely strong correlations emerged between the social bonding variables. The coefficient for "Participation in school activities was important to me" (HSParticipate) and "I tried to stay involved in activities at school" (HSInvolved) for the high school senior time period was .918. Similarly, for the college freshman time period, the coefficient between "Participation in school activities is important to me" (ColParticipate) and "I try to stay involved in activities at school" (ColInvolved) was .906. These strong correlations indicated that, not surprisingly, respondents who felt participation in school activities was important were very likely to stay involved in activities at school.

CORRELATIONS BETWEEN INDEPENDENT VARIABLES AND DEPENDENT VARIABLES

Finally, the last correlations are depictions of the statistically significant correlations between independent variables and dependent variables during the high school time period. As can be seen in the table, there are many significant correlations between the independent and dependent variables, but only a few that are somewhat highly correlated. Overall, the independent variables that were the most highly and consistently correlated with the dependent variables were measures of target suitability.

For example, participation in offline contact with a person met online (PartOffline) was positively associated with providing various types of personal information to online contacts (Person) (r = .405, p < .01), specifically including age (Age) (r = .432, p < .01), gender (Gen) (r = .438, p < .01) and a picture (Pic) (r = .415, p < .01). This suggests that a respondent who provided various types of personal information to an online contact had a higher likelihood of experiencing a relationship offline. With regard to engaging in specific types of offline contact, there were also notable correlations with providing personal information to online contacts. For example, participating in a telephone contact (Telephone) was associated with providing a telephone number to an online contact (Tel) (r = .445, p < .01), and exchange of mail (Mail) was associated with providing information on family conflicts (Fam) (r = .550, p < .01). The independent variables most associated with participation in a sexual encounter (SexEnc) were descriptive characteristics (Des) (r = .477, p < .01) and telephone number (Tel) (r = .448, p < .01). This indicates an increased likelihood of participating in a sexual encounter with an online contact if the respondent provided descriptive characteristics or a telephone number to the online contact.

There were also significant correlations between the independent variables and dependent variables during the college freshman time period. As can be seen in the table, and much like the high school senior time period, there are many significant correlations between the independent and dependent variables, but only a few that are moderately strong. Again, the independent variables that were most highly and consistently correlated with the dependent variables were measures of target suitability.

For example, participation in offline contact with a person met online (PartOffline) was associated with providing personal information to online contacts (Person) (r = .413, p < .01), including age (Age) (r = .401, p < .01) and gender (Gen) (r = .412, p < .01). Again, this indicates that a college freshman who provided personal information to an online contact had a higher likelihood of experiencing a relationship offline. With regard to engaging in specific types of offline contact, there were various correlations with providing personal information to online contacts. For example, exchange of mail (Mail) was positively associated with providing information on emotional distresses (Emot) (r = .424, p < .01), as well as family conflict (r = .455, p < .01). A respondent who provided these types of personal information had a higher likelihood of participating in offline correspondence via the U.S. Mail.

SUMMARY

The purpose of this chapter was to provide a detailed description of the univariate and bivariate results from the analysis of the data collected from 483 college freshmen at a mid-sized university in the northeast. Data originated from surveys administered to the respondents during the 2008 spring semester in freshmen-level classes. The survey questions were derived from the theoretical constructs of Routine Activities Theory, and data were collected on the types of Internet behaviors respondents participated in while a high school senior and a college freshman. Noted below is a brief summary of the findings revealed in this chapter.

Univariate Results

Univariate results were separated by time period in question (i.e., high school senior and college freshman), as well as each theoretical construct. This approach provided a presentation of the results of the frequency and descriptive statistics for each variable in the survey. With regard to Internet behaviors categorized under the construct of exposure to motivated offenders, respondents reported involving themselves in various types of behaviors during both time periods. College freshmen reported spending more time on the Internet (4.21 hours per day and 6.49 days per week) compared to high school seniors (2.85 hours per day and 5.92 days per week). Both groups, however, exhibited large percentages who reported using all four methods of CMCs while online.

The second set of survey questions asked respondents about their behaviors that potentially increased or decreased their target suitability while online. For example, two hundred thirty-seven (49.3%) of the students reported using a non-privatized social networking website as a high school senior, potentially making them a more suitable target, as compared to the 244 (50.7%) that either did not use a social networking website or privatized their website. Two hundred (58.5%) college freshmen used a social networking website that was not privatized. Respondents reported posting personal information on their social networking website and providing it to online contacts during both time periods; however, as college freshmen, slightly more respondents reported they communicated with strangers online (n = 221) as compared to when they were high school seniors (n = 207).

Measures representing lack of capable guardianship were next in the survey. These items questioned respondents on the location of their computer use, as well as restrictions and protections they experienced during Internet use. The majority of the respondents reported using the Internet at home as a high school senior (n = 447), while the majority reported using the Internet in their dorm as a college freshman (n = 422). Overall, as high school seniors, they were more likely to have someone else in the room with them during Internet use (especially a parent or sibling), and they were more likely to have restrictions on time spent online and viewing of adult websites. In the same respect, as high school seniors, they were more likely to be actively monitored (n = 67) or have filtering and blocking software on their computers (n = 239), as compared to when they were college freshmen, at which time a smaller amount were actively monitored (n = 13) or used filtering and blocking software (n = 191).

Finally, with regard to online victimization, the students reported experiencing various occurrences while online. During the high school senior time period, approximately 108 (22.8%) reported experiencing the receipt of unwanted sexually explicit material, while only 48 (11%) college freshmen reported experiencing the receipt of unwanted sexually explicit material. A larger number of respondents reported experiencing the receipt of non-sexual harassment, with 144 (30.8%) receiving it at least one time per week as high school seniors and 52 (13%) receiving it at least one time per week as college freshmen. Finally, with regard to solicitation for sex, 45 (9.5%) reported receiving it as high school seniors, while 37 (7.75) received it as college freshmen.

Respondents also were questioned on their experiences with offline relationships with people they met online. Eighty-five respondents as high school seniors (17.7%), versus 79 (16.4%) as college freshmen, participated in offline contact, which included such things as speaking on the telephone or meeting the person at a home or different location. As a result of these offline contacts, 20 (4.2%) as high school seniors and 12 (2.5%) as college freshmen reported participating in a willing sexual relationship with their online contact.

Bivariate Results

The next section of the chapter explored the bivariate correlations among the independent and dependent variables. These coefficients depicted the general associations among the variables. Bivariate results were separated by time period in question (i.e., high school senior and college freshman), as well as each theoretical construct.

First, the variables representing exposure to motivated offenders and their associations with each other were examined. For both time periods, hours per week spent on the Internet was correlated with the time spent on specific forms of CMCs, indicating that there was a higher likelihood of spending more hours per week on CMCs if the person spent more time on the Internet. With regard to college freshmen, a strong correlation was found between the use of social networking websites and the use of Facebook, so it appeared Facebook was the social networking website used most often by college freshmen.

With regard to variables representing target suitability, moderately strong correlations were found for both groups between the variable representing communication with online contacts and the variables representing various types of information posted on a social networking website or given to an online contact (e.g., age, gender, and picture). It was also indicated that respondents were likely to provide more than one type of personal information to an online contact or on a social networking website. For example, high school seniors who provided a picture were more likely to share emotional distresses or sexual information with an online contact. College freshmen who provided a telephone number to an online contact were more likely to discuss emotional distresses or family conflicts.

Variables representing lack of guardianship were the next set of measures evaluated. For both time periods, main use of a home computer was inversely associated with main use at a school computer lab. Specific to high school seniors, there was a positive correlation between computer use at home and the presence of a parent or sibling in the room, which was not the case for college freshmen.

An examination of the dependent variables (i.e., types of victimization and formation of relationships) showed similarities in correlations for each age group. High school seniors and college freshmen exhibited correlations between forming virtual relationships and participation in several types of offline contact, such as telephone calls and meeting an online contact in person. Participating in offline contact was moderately associated with the experience of a willing sexual encounter.

After review of the control variables, the only moderate to strong correlations were between a few of the social bonding measures. Specifically, there was a strong association between the measures representing involvement within both time periods. There also were significant correlations between all three parent-child conflict variables for both time periods, indicating that if a person experienced higher levels of one type of conflict, it was likely the other two types of conflict were experienced as well.

The correlations between the independent/control variables and the dependent variables were examined last. While there were many statistically significant correlations, there were only a few that were moderately strong. These associations all involved relationships between a dependent variable and types of personal information provided to an online contact. This suggests that providing various types of personal information (representing the theoretical construct of target suitability) increased the likelihood of victimization and forming relationships while online.

The next chapter provides a discussion of the multivariate analysis of the data for the high school senior and college freshman time periods. Each of the seven original hypotheses will be tested for each time period with the use of logistic regression models. Split models also will be used to consider differences in predictors between the male and female respondents.

CHAPTER 6:

Uncovering Potential Factors of Online Victimization

In this chapter, logistic regression models are presented to further assess relationships between the independent variables and the dependent variables. The models developed were based on the theoretical constructs of Routine Activities Theory and were utilized to examine the effects of various independent variables on the likelihood of the dependent variables occurring, for respondents both as high school seniors and college freshmen. The models initially were developed and estimated for the sample as a whole, and then further models were utilized to consider possible differences between males and females in the sample.

Due to the large number of independent variables measured in this study, stepwise logistic regression was utilized to build and produce the final models. In multivariate analysis, some variables can have a statistically significant effect only when another variable is controlled, which is called a suppressor effect (Agresti & Finlay, 1997). Therefore, backward elimination was selected as the method of stepwise regression, whereby all possible variables are initially contained in the model, and there is less risk of ruling out variables involved in suppressor effects (Menard, 2002).

Another step taken to enhance the discovery of potential relationships was to relax the $p < .05$ criterion for retention of variables in the models. Bendel and Afifi (1977) asserted that $p < .05$ is too low and further recommended that the criterion for retention in the stepwise model be set at .15 or .20, so important variables are not excluded. Therefore, the criterion for retention of variables in this study was set at .20, to better reveal any possible statistically significant relationships. Furthermore, linear probability models first were utilized to identify any possible problems with multicollinearity, through the use of tolerance statistics and variance inflation factors.

The effects of the independent variables were assessed in the stepwise logistic regression models through a consideration of one theoretical construct at a time, eventually building the final models shown in Tables 20 through 47. First, variables measuring the theoretical construct of exposure to motivated offenders were inserted in the models to examine their effects on the dependent variables. The next model considered the addition of the effects of the independent variables measuring the theoretical construct of target suitability, while also including retained variables measuring exposure to motivated offenders. Third, lack of capable guardianship variables were assessed, in addition to the effects of the other two sets of retained independent variables. Finally, full models were constructed with the addition of the control variables, while also including the retained measures of the three theoretical constructs. The results of these analyses are presented in the remainder of this chapter.

LOGISTIC REGRESSION MODELS

High School Senior Time Period

Full Models. Table 20 presents the logistic regression estimates for "receipt of unwanted sexually explicit material" during the high school senior time period. The first block of the model initially included the entire set of independent variables representing the theoretical construct of exposure to motivated offenders (IntHours, Research, Gaming, Travel, Design, Shop, Social, Other, EmHours, IMHours, ChatHour, SNWHours, MySpace, and Facebook). Variables retained at the .20 level were shown to explain only 5.6% to 8.4% of the variation in the dependent variable. Three variables emerged as statistically significant predictors ($p < .05$) of the dependent variable.

Respondents who shopped online (Shop) were significantly more likely to receive unwanted sexually explicit material ($b = .641$, $p < .01$), along with respondents who performed activities designated as "Other" (most often noted as "banking") ($b = .651$, $p < .05$). In addition, according to this partial model, spending one or more hours per week in chat rooms ($b = 1.218$, $p < .01$) increased the likelihood of receipt of unwanted sexually explicit material more than any other variable. In fact,

Table 20. Logistic Regression Estimates for the Dependent Variable of Receipt of Sexually Explicit Material During the High School Time Period (N = 483)

Variable	Exposure to Motivated Offenders		Target Suitability		Lack of Capable Guardianship		Control Variables	
	B(SE)	Exp(B)	B(SE)	Exp(B)	B(SE)	Exp(B)	B(SE)	Exp(B)
Travel	-.375(.252)	.687	-.369(.253)	.692	-.454(.261)	.635	-.456(.269)	.634
Design	.485(.285)	1.624	.445(.289)	1.561	.516(.295)	1.676	.477(.305)	1.611
Shop	.641(.245)	1.898**	.624(.246)	1.869*	.732(.256)	2.079**	.812(.263)	2.253**
Social	.611(.439)	1.841	—	—	—	—	—	—
Other	.651(.329)	1.917*	.630(.326)	1.878	.640(.337)	1.896	.592(.357)	1.808
ChatHour	1.218(.357)	3.379**	.971(.375)	2.642*	.959(.385)	2.608*	.774(.393)	2.169*
ProvidedInfo	.110(.042)	1.115*	.116(.044)	1.122**	.107(.046)	1.113*		
SchLab			.822(.521)	2.275	—	—		
ParInRm			.400(.232)	1.491	.446(.237)	1.562		
OthInRm			.512(.345)	1.714	.489(.357)	1.630		
RestrictTime			.539(.296)	1.699	.498(.303)	1.645		
Sex					-.436(.242)	.646		
White					-.750(.297)	.472*		
GPA					-.222(.109)	.801		
Privileges					.142(.043)	1.153**		
Constant	-2.259(.454)	.104***	-1.814(.215)	.163***	-2.270(.272)	.103***	-1.464(.440)	.231**
-2 Log-likelihood	502.553		498.252		486.460		467.669	
Model Chi-Square	27.766***		32.067**		43.860***		62.651	
Cox & Snell R^2	056		065		087		123	
Nagelkerke R^2	084		097		131		183	

137

respondents who reported using chat rooms an hour or more per week (ChatHour) were over three times more likely to receive sexually explicit material [Exp(B) = 3.379].

The second set of independent variables, representing the theoretical construct of target suitability (NoPriv, SNWInfo, Comm, and ProvidedInfo), added another to the list of significant predictors of victimization through the receipt of unwanted sexually explicit material. In order to deal with multicollinearity among variables representing the types of information that could be posted on a social networking website and provided to online contacts, the variables "SNWInfo" (α = .796) and "ProvidedInfo" (α = .911) were created as additive measures of the original dichotomous variables. This model was shown to explain a limited range of 6.5% to 9.7% of the variation in the dependent variable. After the addition of target suitability variables, shopping online (Shop) (b = .624, p <.05) and use of chat rooms (ChatHour) (b = .971, p < .05) continued to significantly increase the likelihood of receipt of unwanted sexually explicit material. Respondents who provided various types of personal information to online contacts were also more likely to receive unwanted sexually explicit material (b = .110, p < .05).

Addition of the third set of independent variables, representing the theoretical construct of lack of capable guardianship (LivRm, SchLab, FriHom, OthRm, ParInRm, FriInRm, TeachInRm, SibInRm, OthInRm, NoOneRm, RestrictTime, RestrictAdult, RestrictCMC, RestrictOther, ActMon, DKActMon, FiltSoft, DKFiltSoft) did not produce the emergence of any further statistically significant variables (p < .05). Variables retained at the .20 level were shown to explain 8.7% to 13.1% of the variation in the dependent variable. Shopping online (Shop) did experience an increase in its significance as a predictor of receipt of unwanted sexually explicit material (b = .732, p < .01); it was shown that respondents who shopped online were two times more likely to be victimized in this manner [Exp(B) = 2.079]. Although the significance of use of chat rooms (ChatHour) decreased slightly, it was still the case that respondents who participated in chat rooms were over 2.5 times more likely to receive unwanted sexually explicit material [Exp(B) = 2.608]. Finally, providing various types of personal information to online contacts (ProvidedInfo) continued to significantly increase the likelihood of receipt of unwanted sexually explicit material (b = .116, p < .01).

Control variables (Sex, Age, White, GPA, ShareParents, ShareFriends, EnjoyFriends, RespectParents, RespectTeachers, Participate, Grades, Succeed, Yell, Nag, and Privileges) were added last to create the full logistic regression model presented in Table 20. The full model was shown to explain a range of 12.3% to 18.3% of the variation in the dependent variable. Similar to the partial models, participation in the behaviors represented by the three previously mentioned significant variables continued to increase the likelihood of receipt of unwanted sexually explicit material. Respondents who shopped online (Shop) and those who used chat rooms one or more hours per week (Chat) were over two times more likely to be victimized, and those who provided various types of information to online contacts also were more likely to receive unwanted sexual material. In addition, two control variables emerged as significant predictors. First, respondents who were white (White) were less likely than nonwhites to receive unwanted sexually explicit material online (b = -.750, p < .05). Second, respondents whose parents more often took away privileges (Privileges) during the high school senior time period were more likely to be victimized (b = .142 p < .01). The temporal ordering of the latter relationship may be important to consider, as it is possible that when respondents received sexually explicit material, parents then took away privileges.

Table 21 presents the logistic regression estimates for the dependent variable "receipt of non-sexual harassment" during the high school time period. After the insertion of the first full set of independent variables, representing the theoretical construct of exposure to motivated offenders, two variables emerged as significant predictors of the dependent variable: socializing while online and hours per week spent using instant messaging. Moreover, variables retained at the .20 level were shown to explain 7.7% to 10.8% of the variation in the dependent variable. Respondents who socialized online (Social) were over 5 times more likely to receive non-sexual harassment [Exp(B) = 5.414]. Also, according to the partial model, greater hours per week spent in instant messaging (IMHours) also significantly increased the likelihood of receipt of non-sexual harassment (b = .058, p < .05).

Table 21. Logistic Regression Estimates for the Dependent Variable of Receipt of Non-Sexual Harassment During the High School Senior Time Period (N = 483)

Variable	Exposure to Motivated Offenders B(SE)	Exp(B)	Target Suitability B(SE)	Exp(B)	Lack of Capable Guardianship B(SE)	Exp(B)	Control Variables B(SE)	Exp(B)
IntHours	.017(.013)	1.017	—	—	—	—	—	—
Design	.484(.262)	1.622	.392(.270)	1.479	.421(.275)	1.524	—	—
Shop	.311(.213)	1.365	.339(.214)	1.404	.336(.220)	1.400	.344(.221)	1.410
Social	1.689(.617)	5.414**	1.564(.617)	4.778*	1.732(.632)	5.655**	1.537(.631)	4.651*
EmHours	.141(.106)	1.151	.179(.106)	1.196	.166(.109)	1.181	.232(.111)	1.261*
IMHours	.058(.027)	1.060*	.058(.027)	1.059*	.050(.028)	1.051	.048(.028)	1.049
ProvidedInfo	—	—	.168(.041)	1.183***	.160(.042)	1.173***	.178(.043)	1.195***
LivRm					-.453(.239)	.636	-.391(.221)	.677
SchLab					-1.172(.819)	.310	-1.420(.809)	.242
ParInRm					.373(.256)	1.452	—	—
NoOneRm					.522(.246)	1.686*	.767(.459)	2.152
RestrictCMC					.859(.460)	2.362	-.331(.256)	.718
DKActMon					-.331(.231)	.718	.381(.240)	1.464
Sex							-.181(.112)	.835
GPA							.154(.089)	1.167
Grades							-.184(.075)	.832*
Succeed							—	—
Constant	-3.310(.637)	.037***	-3.198(.629)	.041***	-3.465(.685)	.031***	-2.707(.872)	.067**
-2 Log-likelihood	568.840		552.591		537.341		525.572	
Model Chi-Square	38.762***		55.011***		70.260***		82.029***	
Cox & Snell R^2	.077		.108		.136		.157	
Nagelkerke R^2	.108		.151		.189		.219	

$* \ p < .05, \ ** \ p < .01, \ *** \ p < .00$

The second full set of independent variables, representing the theoretical construct of target suitability, added a third variable to the list of significant predictors of victimization through receipt of non-sexual harassment. Variables retained at the .20 level in this model were shown to explain 10.8% to 15.1% of the variation in the dependent variable. After the addition of target suitability variables, socializing online (Social) (b = 1.564, p < .05) continued to have a significant effect on victimization; respondents who socialized online still were nearly 5 times more likely to receive non-sexual harassment [Exp (B) = 4.778]. The effect of spending greater hours per week using instant messaging (IMHours) also remained the same (b = .058, p < .05). Finally, respondents who provided various types of information to online contacts (ProvidedInfo) were also more likely to be victimized online through the receipt of non-sexual harassment (b = .168, p < .001).

Three variables emerged as significant predictors of victimization after the addition of the third set of independent variables, representing the theoretical construct of lack of capable guardianship. Variables retained at the .20 level now were shown to explain 13.6% to 18.9% of the variation in the dependent variable. Socializing online (Social) continued to significantly increase the likelihood of non-sexual harassment (b = 1.732, p < .01), as respondents who socialized online (Social) were over 5.5 times more likely to be victimized in this manner [Exp (B) = 5.655]. Providing information to online contacts (ProvidedInfo) also continued to increase the likelihood of receipt of non-sexual harassment, by approximately 17% [Exp(B) = 1.173] for each type of information provided. In addition to these two variables, the presence of no other persons in the room (NoOneRm) during Internet use significantly increased the likelihood of receipt of non-sexual harassment (b = .522, p < .001).

Finally, control variables were added to create the full logistic regression model presented in Table 21. Variables retained at the .20 level were shown to explain 15.7% to 21.9% of the variation in the dependent variable. Socializing online (Social) continued to increase the likelihood of non-sexual harassment (b = 1.537, p < .05). Furthermore, hours per week spent using email (EmHours) now emerged as a variable that significantly increased the likelihood of this type of victimization (b = .232, p < .05). Providing various types of personal information to online contacts (ProvidedInfo) was the most statistically significant predictor of non-sexual harassment (b = .178, p < .001). Finally, the only significant control variable in the model was placing an importance on succeeding in school. Respondents who had a stronger desire to succeed in school (Succeed) were less likely to receive non-sexual harassment (b = -.184, p < .05).

The logistic regression estimates for the dependent variable "receipt of sexual solicitation" during the high school time period are presented in Table 22. Exposure to motivated offenders variables retained at the .20 level again were shown to explain a limited amount of only 3.7% to 7.3% of the variation in the dependent variable. Use of chat rooms (ChatHour) (b = 1.113, p < .01) significantly increased the likelihood of receipt of sexual solicitation; respondents who participated in this activity were over 3 times more likely to receive sexual solicitation online [Exp (B) = 3.044]. Use of social networking websites (SNWHours) did not have as substantial of an impact as chat room use (b = .077, p < .05), but still was a significant predictor in increasing the likelihood of receiving sexual solicitation.

Insertion of the independent variables representing the theoretical construct of target suitability decreased the total number of significant variables to one: providing information to online contacts. Variables retained at the .20 level were shown to explain 9.0% to 17.5% of the variation in the dependent variable. Providing personal information to online contacts (ProvidedInfo) was revealed to increase the likelihood of receipt of sexual solicitation by an average of 32% [Exp (B) = 1.319] for each type of personal information provided. As indicated in previous models, participating in this behavior was also shown to increase the likelihood of receipt of sexual material and non-sexual harassment.

A second variable emerged as a significant predictor of victimization after the addition of the third set of independent variables, representing the theoretical construct of lack of capable guardianship. Variables retained at the .20 level in this model were shown to explain 11.3% to 22.0% of the variation in the dependent variable. Providing personal information to online contacts (ProvidedInfo) remained a significant predictor of receiving sexual solicitation online (b = .281, p < .001). Furthermore, through the addition of the guardianship variables, gaming while online (Gaming) significantly decreased the likelihood of sexual solicitation (b = -.633, p < .05).

Table 22. Logistic Regression Estimates for the Dependent Variable of Receipt of Sexual Solicitation During the High School Senior Time Period (N = 483)

Variable	Exposure to Motivated Offenders		Target Suitability		Lack of Capable Guardianship		Control Variables	
	B(SE)	Exp(B)	B(SE)	Exp(B)	B(SE)	Exp(B)	B(SE)	Exp(B)
Gaming	-.393(.294)	.675	-.563(.213).	569	-.633(.321)	.531*	-.567(.343)	.567
Social	1.507(1.036)	4.512	1.164(1.014)	3.203	1.149(1.049)	3.154	1.366(1.089)	3.919
ChatHour	1.113(.410)	3.044**	—	—	—	—	—	—
SNWHours	.077(.038)	1.081*	.064(.041)	1.067	.074(.043)	1.077	.081(.044)	1.085
ProvidedInfo			.277(.047)	1.319***	.281(.049)	1.325***	.320(.055)	1.377***
OthPl					1.920(.817)	6.823	2.196(.949)	8.988*
FriInRm					.474(.327)	1.606	.605(.349)	1.831
SibInRm					-.535(.331)	.586	-.630(.347)	.532
RestrictTime					-.705(.490)	.494	-.761(.511)	.467
ShareFriends							-.228(.089)	.796*
RespectParents							.203(.125)	1.225
RespectTeachers							-.214(.098)	.080*
Succeed							.146(.082)	1.157
Nag							-.122(.072)	.885
Privileges							.143(.071)	1.153*
Constant	-3.739(1.020)	.024***	-3.663(1.018)	.026***	-3.690(1.033)	.025***	-3.214(1.484)	.040*
-2 Log-likelihood	327.819		300.730		288.508		265.676	
Model Chi-Square	18.250**		45.339***		57.561***		80.393***	
Cox & Snell R²	.037		.090		.113		.154	
Nagelkerke R²	.073		.175		.220		.300	

* p < .05   ** p < .01   *** p < .001

The full logistic regression model, containing the control variables, is presented in the final column of Table 22. Variables retained at the .20 level were found to explain 15.4% to 30.0% of the variation in the dependent variable. Two independent variables were statistically significant predictors, along with three control variables. Providing personal information to online contacts (ProvidedInfo) had the most highly significant impact on this type of victimization, as it increased the likelihood of receipt of sexual solicitation by approximately 38% [Exp(B) = 1.377] for each type of information provided. Main use of the Internet in locations noted as "Other Place" (OtherPl) (i.e., not in the parent's or friend's home, or school computer lab) also significantly increased this likelihood (b = 2.196, p < .05). With regard to the control variables, respondents who reported they could share thoughts and feelings with friends (ShareFriends) (b = -.228, p < .05) and those who had greater respect for their teachers (RespectTeachers) (b = -.214, p < .05) were significantly less likely to receive sexual solicitation online. In addition, a respondent whose parents more often took away privileges (Privileges) was more likely to receive sexual solicitation online (b = .143, p < .05). Again, this could indicate that privileges were removed as a result of inappropriate online behaviors, or that conflict with parents actually impacted the likelihood that a respondent would receive sexual solicitation online.

The remaining dependent variables assessed in this study focused on the formation of various types of relationships with online contacts. Logistic regression estimates for the dependent variable "formation of a virtual relationship with an online contact" during the high school senior time period are presented in Table 23. Multiple variables representing the theoretical construct of exposure to motivated offenders emerged as significant predictors of the dependent variable. Variables retained at the .20 level were shown to explain 9.2% to 14.7% of the variation in this dependent variable.

Increased use of various types of CMCs influenced the likelihood of forming a virtual relationship with an online contact. Respondents who exhibited greater use of instant messaging (IMHours) (b = .079, p < .05) had a significantly greater likelihood of forming a relationship online. Much like was seen in previous models, use of chat rooms

Table 23. Logistic Regression Estimates for the Dependent Variable Formation of Virtual Relationship During the High School Senior Time Period (N = 483)

Variable	Exposure to Motivated Offenders		Target Suitability		Lack of Capable Guardianship		Control Variables	
	B(SE)	Exp(B)	B(SE)	Exp(B)	B(SE)	Exp(B)	B(SE)	Exp(B)
Social	1.137(.764)	3.117		—		—		—
IMHours	.079(.034)	1.082*	.087(.039)	1.204*	.091(.042)	1.096*	.099(.043)	1.104*
ChatHour	1.388(.366)	4.007***	.675(.419)	1.964	.700(.436)	2.015	.615(.440)	1.849
SNWHours	-.093(.043)	.912*	-.139(.049)	.870**	-.150(.051)	.860**	-.151(.052)	.860**
MySpace	1.087(.318)	2.967*	.832(.346)	2.297*	.830(.359)	2.293*	.936(.371)	2.550*
Facebook	.445(.279)	1.560	.634(.307)	1.886*	.849(.331)	2.337*	.899(.340)	2.457**
Comm			1.656(.325)	5.239***	1.865(.346)	3.467***	1.875(.374)	6.520***
ProvidedInfo			.181(.049)	1.199***	.193(.051)	1.213***	.208(.053)	1.231***
SchLab					1.565(.789)	4.781*	1.640(.821)	5.154*
FrInRm					-.630(.310)	.533*	-.657(.319)	.519*
SibInRm					.525(.326)	1.691	.494(.332)	1.639
NoOneRm					1.148(.331)	3.152**	1.180(.345)	3.256*
DKFiltSoft					1.030(.443)	2.800*	.980(.447)	2.665*
White							-.552(.374)	.593
ShareParents							.106(.064)	1.112
RespectParents							-.179(.090)	.836*
Constant	-3.697(.762)	.025***	-3.626(.418)	.027***	-4.591(.566)	.010***	-3.405(.998)	.033**
-2 Log-likelihood	428.778		362.447		338.839		329.066	
Model Chi-Square	46.446***		112.778***		136.386***		146.158***	
Cox & Snell R^2	.092		.209		.247		.262	
Nagelkerke R^2	.147		.333		.393		.417	

* $p < .05$, ** $p < .01$, *** $p < .001$ (b = 1.656, p < .001)

145

(ChatHour) had the largest impact, as these respondents were 4 times more likely to form a virtual relationship [Exp(B) = 4.007]. Increased use of social networking websites (SNWHours) actually decreased the likelihood of a virtual relationship (b = -.093, p < .05). On the other hand, reported use of a specific type of social networking website, MySpace, significantly increased the likelihood of a virtual relationship (b = 1.087, p < .05).

Additional significant predictors that affected the likelihood of forming a virtual relationship online were present after the insertion of variables representing the theoretical construct of target suitability. In this model, variables retained at the .20 level were shown to explain 20.9% to 33.3% of the variation in the dependent variable. Greater hours per week of instant messaging (IMHours) (b = .832, p < .05) still was shown to increase the likelihood of formation of a virtual relationship. Both the use of MySpace (b = .832, p < .05) and Facebook (b = .634, p < .05) significantly increased the likelihood of the dependent variable, but greater hours per week use of social networking websites was shown to decrease the likelihood of forming a virtual relationship (b = -.139, p < .01).

With regard to the target suitability variables, both communicating with people online (Comm) and providing personal information to online contacts (ProvidedInfo) (b = .181, p < .001) had highly significant positive impacts on the likelihood of the dependent variable. In particular, respondents who communicated with people online (Comm) were over 5 times more likely to form a virtual relationship with an online contact [Exp(B) = 5.239].

Other variables emerged as significant predictors after the addition of the third set of independent variables, representing the theoretical construct of lack of capable guardianship. The variables retained at the .20 level were shown to explain 24.7% to 39.3% of the variation in the dependent variable. Greater hours per week of instant messaging (IMHours) (b = .091, p < .05), use of MySpace (b = .830, p < .01), and use of Facebook (b = .849, p < .05) continued to increase the likelihood of forming a virtual relationship, while greater hours per week use of social networking websites (b = -.150, p < .05) continued to decrease this likelihood. Communicating with others online (Comm) (b = 1.865, p < .001) and providing personal information to online contacts (ProvidedInfo) (b = .193, p < .001) maintained their position as highly significant predictors of forming a virtual relationship with an online contact. Furthermore, several guardianship variables also influenced the occurrence of the dependent variable. Using a school computer lab as the main location of Internet use (SchLab) increased the likelihood, (b = 1.565, p < .05), as did the respondent being unsure if filtering and blocking software was present on the computer (DKFiltSoft) (b = 1.030, p < .05). The presence of a friend in the room during Internet use (FriInRm) decreased the likelihood of forming a virtual relationship online (b = -.630, p < .05), while having no friend or family member in the room (NoOneRm) increased the likelihood (by more than 3 times) of the dependent variable occurring [Exp (B) = 3.152].

Finally, control variables were added to create the full logistic regression model presented in Table 23. Variables retained at the .20 level were shown to explain a respectable 26.2% to 41.7% of the variation in the dependent variable. The same variables mentioned previously remained as significant predictors that affected the likelihood of the dependent variable occurring, with communicating with people online (Comm) (b = 1.875, p < .001) and providing personal information to online contacts (ProvidedInfo) (b = .208, p < .001) emerging as the most highly significant predictors of forming a virtual relationship. One control variable, respect for parents (RespectParents), also was significant. Respondents who reported greater respect for their parents were less likely to form a virtual relationship with an online contact (b = -.179, p < .05).

Table 24 presents the logistic regression estimates for the dependent variable "participation in offline contact" with a person met online during the high school time period. After the insertion of the first set of independent variables, representing the theoretical construct of exposure to motivated offenders, two variables emerged as significant predictors. Variables retained at the .20 level also were shown to explain 9.0% to 14.9% of the variation in the dependent variable. Much like was seen in previous models, use of chat rooms (ChatHour) had a large impact, as these respondents were over 2.5 times more likely to participate in offline contact [Exp(B) = 2.684]. Use of MySpace also similarly increased the likelihood of participation in offline contact with a person met online [Exp(B) = 2.811].

Table 24. Logistic Regression Estimates for the Dependent Variable of Participation in Offline Contact During the High School Senior Time Period (N = 483)

Variable	Exposure to Motivated Offenders		Target Suitability		Lack of Capable Guardianship		Control Variables	
	B(SE)	Exp(B)	B(SE)	Exp(B)	B(SE)	Exp(B)	B(SE)	Exp(B)
Research	1.358(1.072)	3.887	—	—	—	—	—	—
EmHours	-.243(.139)	.784	-.330(.154)	.719*	-.368(.160)	.692*	-.409(.171)	.665*
ChatHour	.987(.379)	2.684**	—	—	—	—	—	—
SNWHours	.049(.036)	1.051	.074(.045)	1.077	.077(.047)	1.080	.101(.049)	1.106*
MySpace	1.033(.340)	2.811**	.814(.434)	2.257	1.032(.447)	2.806*	1.042(.480)	2.836*
NoPrivSNW			-.711(.321)	.491*	-.769(.335)	.464*	-1.106(.360)	.362**
SNWInfo			-.156(.086)	.855	-.178(.069)	.837*	-.202(.094)	.817*
Comm			1.922(.379)	6.834***	1.911(.386)	6.760***	2.028(.404)	7.600***
ProvidedInfo			.296(.054)	1.345***	.297(.056)	1.346***	.311(.059)	1.365***
OthPl					2.024(.938)	7.567*	2.133(.942)	8.443*
RestrictAdult					-.914(.352)	.401**	-1.098(.372)	.334**
Sex							-1.022(.344)	.360**
ShareParents							.137(.069)	1.147*
EnjoyFriends							-.176(.106)	.839
RespectParents							-.197(.094)	.821*
Succeed							.149(.071)	1.161*
Constant	-5.529(1.475)	.004***	-2.788(.501)	.062***	-2.570(.516)	.077***	-.779(1.099)	.459
-2 Log-likelihood	403.231		312.265		298.137		281.063	
Model Chi-Square	45.422***		136.389***		150.517***		167.591***	
Cox & Snell R²	.090		.247		.269		.294	
Nagelkerke R²	.149		.407		.433		.485	

* $p < .05$, ** $p < .01$, *** $p < .001$

The second set of independent variables, representing the theoretical construct of target suitability, affected the number and types of variables that influenced the likelihood of participation in offline contact. Variables retained at the .20 level were shown to explain 24.7% to 40.7% of the variation in the dependent variable. Greater hours per week use of email (b = -.330, p < .05) emerged as a factor that actually decreased the likelihood of the dependent variable occurring, as did use of a non-privatized social networking website (b = -.711, p < .01). With regard to other target suitability variables, both communicating with people online (Comm) (b = 1.922, p < .001) and providing personal information to online contacts (ProvidedInfo) (b = .296, p < .001) had highly significant impacts on the likelihood of the dependent variable. Respondents who reported communicating with people online were almost 7 times more likely to participate in offline contact with a person met online [Exp(B) = 6.834].

Further variables emerged as significant predictors after the addition of the third set of independent variables, representing the theoretical construct of lack of capable guardianship. Variables retained at the .20 level were shown to explain 26.9% to 43.3% of the variation in the dependent variable. Increased hours per week of email use (EmHours) (b = -.368, p < .05) and using a non-privatized social networking website (NoPriv) (b = -.769, p < .05) continued to decrease the likelihood of participating in offline contact, as did providing personal information on a social networking website (SNWInfo) (b = -.178, p < .05).

Communicating with others online (Comm) (b = 1.911, p < .001) and providing personal information to online contacts (ProvidedInfo) (b = .297, p < .001) maintained their position as highly significant predictors of participating in offline contact with a person met online. In addition, two guardianship variables also influenced the dependent variable. Using a location designated as "Other" (most commonly noted as boyfriend or girlfriend's home) as the main location of Internet use (OtherPl) increased the likelihood of the dependent variable occurring (b = 2.024, p < .05), while respondents who had restrictions on viewing adult websites (RestrictAdult) experienced a decrease in the likelihood of participating in offline contact (b = -.914, p < .01).

Addition of the control variables created the full logistic regression model presented in the final column of Table 24. All variables retained at the .20 level were shown to explain a healthy 29.4% to 48.5% of the variation in the dependent variable. Variables mentioned previously remained as significant predictors that affected the likelihood of participating in offline contact, with communicating with people online (Comm) (b = 2.028, p < .001) and providing personal information to online contacts (ProvidedInfo) (b = .311, p < .001) remaining the most highly significant predictors of the dependent variable. Several control variables also were significant. Females were less likely than males to participate in offline contact (Sex) (b = -1.022, p < .01), as were respondents who reported greater respect for their parents (RespectParents) (b = -.197, p < .05). On the other hand, respondents who more often shared their thoughts and feelings with their parents (ShareParents) (b = .137, p < .05) and those who tried hard to succeed in school (Succeed) (b = .149, p < .05) were more likely to participate in offline contact with a person met online.

Logistic regression estimates for the dependent variable "participation in face-to-face contact with a person met online" during the high school senior time period are presented in Table 25. Respondents were asked in the survey to note what types of offline contact they had participated in with a person met online, and three of those choices included contact that involved a face-to-face meeting: met at person's home, met at your home, and met at another location. If a respondent participated in one or more of these meetings, they were coded as participating in face-to-face contact. After the insertion of the first set of independent variables, representing the theoretical construct of exposure to motivated offenders, only one variable was a significant predictor of the dependent variable.

Also, variables retained at the .20 level were shown to explain only 5.6% to 10.9% of the variation in the dependent variable. Use of MySpace as a social networking website significantly increased the likelihood of participating in a face- to-face contact with a person met online (b = 1.682, p < .001). Respondents who used MySpace were over 5 times more likely to participate in this behavior [Exp(B) = 5.379].

The second set of independent variables, representing the theoretical construct of target suitability, increased the number of vari-ables to three that were significant predictors of face-to-face contact.

Table 25. Logistic Regression Estimates for the Dependent Variable Face-to-Face Contact During the High School Senior Time Period (N = 483)

Variable	Exposure to Motivated Offenders		Target Suitability		Lack of Capable Guardianship		Control Variables	
	B(SE)	Exp(B)	B(SE)	Exp(B)	B(SE)	Exp(B)	B(SE)	Exp(B)
Other	.622(.415)	1.863	.849(.458)	2.338	.951(.473)	2.588*	.896(.497)	2.450
MySpace	1.682(.452)	5.379***	1.125(.493)	3.808*	1.231(.500)	3.426*	1.287(.523)	3.622*
Facebook	.528(.342)	1.696	—	—	—	—	—	
NoPrivSNW			-.561(.344)	.570	-.560(.364)	.571	-.713(.381)	.490
Comm			1.446(.441)	4.426***	1.427(.452)	4.168**	1.395(.460)	4.037**
ProvidedInfo			.271(.054)	1.311***	.276(.056)	1.318	.295(.060)	1.343***
LivRm					.476(.346)	1.610	.464(.358)	1.590
OthPl					1.829(.983)	6.230	1.941(.968)	6.964*
RestrictAdult					-1.153(.436)	.316**	-1.262(.450)	.283**
RestrictCMC					1.461(.650)	4.309*	1.576(.675)	4.836*
ActMon					.652(.465)	1.920	.690(.485)	1.993
Sex							-.599(.397)	.549
LivDorm							.885(.536)	2.423
ShareParents							.119(.077)	1.126
RespectParents							-.192(.103)	.825
Constant	-3.748(.469)	.024***	-4.260(.569)	.014***	-4.498(.603)	.011	-4.120(1.016)	.016***
-2 Log-likelihood	322.740		259.032		244.299		234.885	
Model Chi-Square	27.866***		91.574***		106.307***		115.271***	
Cox & Snell R²	.056		.173		.198		.213	
Nagelkerke R²	.109		.335		.383		.413	

* p < .05; ** p < .01; *** p < .001

151

Furthermore, variables retained at the .20 level were shown to explain 17.3% to 33.5% of the variation in the dependent variable. The use of MySpace remained significant (b = 1.125, p < .05). With regard to the target suitability variables, similar to previous models, both communicating with people online (Comm) (b = 1.446, p < .001) and providing various types of personal information to online contacts (ProvidedInfo) (b = .271, p < .001) had highly significant effects on the likelihood of a face-to-face contact. Respondents who reported communicating with people online were over 4 times more likely to participate in face-to-face contact with a person met online [Exp(B) = 4.426].

Other variables emerged as significant predictors of the dependent variable after the addition of the third set of independent variables, representing the theoretical construct of lack of capable guardianship. Variables retained at the .20 level in this model were shown to explain 19.8% to 38.3% of the variation in the dependent variable. Performing activities marked as "Other" while online (Other) (most commonly noted as "banking") (b = .951, p < .05) and use of MySpace (b = 1.231) both increased the likelihood of participating in offline contact with a person met online. Communicating with others online (Comm) (b = 1.427, p < .01) maintained its position as a highly significant predictor, but providing information to online contacts (ProvidedInfo) lost its significance. Two guardianship variables also influenced the occurrence of the dependent variable. Respondents who had restrictions on viewing adult websites (RestrictAdult) experienced a lower likelihood of participating in offline face-to-face contact (b = -1.153, p < .01), while respondents who had restrictions on the use of CMCs (RestrictCMC) were significantly more likely to participate in the behavior (b = 1.461, p < .05).

The full logistic regression model, after the addition of the control variables, is presented in Table 25. Variables retained at the .20 level were shown to explain a respectable 21.3% to 41.3% of the variation in the dependent variable. Use of MySpace (b = 1.287, p < .05), communicating with people online (Comm) (b = 1.395, p < .01), and providing personal information to online contacts (ProvidedInfo) (b = .295, p < .001) emerged as significant predictors. Respondents whose main Internet use was in a location designated as "Other" (OtherPl) (most commonly noted as the home of a boyfriend or girlfriend) (b = 1.941, p < .05), as well as those who had restrictions on their use of CMCs

(RestrictCMC) (b = 1.576, p < .05) also were more likely to participate in face-to-face contact with people met online. The only factor that was significant in decreasing participation in this behavior was the restriction of adult website viewing (RestrictAdult) (b = -1.262, p < .01), indicating that restrictions on these websites lowered the likelihood a respondent would physically meet with an online contact.

The last set of models examined for the entire sample during the high school senior time period appear in Table 26, which reveals the logistic regression estimates for the dependent variable "participation in a sexual encounter" with an online contact. After the insertion of the first set of independent variables, representing the theoretical construct of exposure to motivated offenders, no variables emerged as statistically significant predictors of this behavior. Moreover, variables retained at the .20 level were shown to explain only 1.8% to 6.3% of the variation in the dependent variable. Variables representing the theoretical construct of target suitability produced one significant predictor, and those variables retained at the .20 level were shown to explain 11.4% to 39.0% of the variation in the dependent variable. Respondents who provided personal information to online contacts (ProvidedInfo) were 56% more likely to participate in a sexual encounter for each type of personal information provided [Exp(B) = 1.566].

A second variable emerged as a significant predictor after the addition of the third set of independent variables, representing the theoretical construct of lack of capable guardianship. Variables retained at the .20 level were shown to explain 12.0% to 41.0% of the variation in the dependent variable. Providing personal information to online contacts (ProvidedInfo) remained statistically significant (b = .488, p < .001). Additionally, shopping while online was shown to decrease the likelihood of participating in a sexual encounter with an online contact (b = -1.234, p < .05).

Finally, control variables were added to create the final logistic regression model presented in Table 26. Variables retained at the .20 level were shown to explain 15.6% to 53.2% of the variation in the dependent variable. Shopping online (Shop) remained a significant negative predictor of the behavior (b = -2.035, p < .01). Providing information to online contacts (ProvidedInfo) still was shown to significantly increase the likelihood of a sexual encounter (b = .743, p < .001).

Table 26. Logistic Regression Estimates for the Dependent Variable Sexual Encounter During the High School Senior Time Period

(N = 483)

Variable	Exposure to Motivated Offenders		Target Suitability		Lack of Capable Guardianship		Control Variables	
	B(SE)	Exp(B)	B(SE)	Exp(B)	B(SE)	Exp(B)	B(SE)	Exp(B)
IntHours	.037(.026)	1.037	—	—	—	—	—	—
Gaming	.736(.504)	2.088	—	—	—	—	—	—
Shop	-.722(.471)	.486	-1.236(.576)	.290	-1.234(.581)	.291*	-2.035(.744)	.131**
Facebook	.956(.576)	2.600	—	—	—	—	—	—
Comm			1.748(1.116)	5.741	1.692(1.119)	5.428	—	—
ProvidedInfo			.448(.094)	1.566***	.488(.102)	1.626***	.743(.143)	2.166***
LivRm					.966(.549)	2.627	1.189(.647)	3.284
ShareParents							.400 (.139)	1.492**
RespectParents							-.699(.188)	.497***
Privileges							.202(.095)	1.224*
Constant	-4.478(.746)	.011	-5.283(1.022)	.005***	-5.852(1.101)	.003***	-3.540(1.097)	.029**
-2 Log-likelihood	157.408		108.101		105.015		85.019	
Model Chi-Square	8.954		58.181***		63.147***		81.343***	
Cox & Snell R^2	.018		.114		.120		.156	
Nagelkerke R^2	.063		.390		.410		.532[a]	

$* p < .05$, $** p < .01$, $*** p < .001$

[a] The Nagelkerke R^2 estimates in these models are quite inflated, as well in other models in this study with relatively little variation in the dependent variable. Along with the impact of limited variation in the dependent variable, the Nagelkerke R^2 tends to be a more liberal estimate of explained variation (Meyers et al., 2006)

154

Three control variables also emerged as significant predictors of a sexual encounter. Respondents who more often shared thoughts and feelings with parents (ShareParents) [Exp(B) = 1.492] and more often had privileges taken away by parents (Privileges) [Exp(B) = 1.224] were more likely to participate in a sexual encounter with an online contact. On the other hand, respondents who reported greater respect for parents (RespectParents) were significantly less likely to participate in a sexual encounter with an online contact (b = -.699, p < .001).

SPLIT MODELS

Table 27 presents the logistic regression estimates for males and females the dependent variable "receipt of unwanted sexually explicit material" during the high school senior time period. Only the full models are presented in the table; stepwise regression with backward elimination was used in the analysis, with the entire set of independent and control variables initially inserted. Variables retained at the .20 level were shown to explain a respectable 29.7% to 42.7% of the variation in the dependent variable for males, but only 11.3% to 17.6% for females. Furthermore, males and females had no common statistically significant predictors. Males in the sample did have a number of significant predictors. For example, use of chat rooms (ChatHour) (b = 2.012, p < .001) and using a school computer lab as the main location of Internet use (SchLab) (b = 2.355, p < .05) substantially increased the likelihood that the males in the sample would receive sexual material. Alternatively, males who had restrictions on viewing adult websites (RestrictAdult) were less likely to receive this material (b = -1.384, p < .01), while those who were unaware if filtering and blocking software was installed on their computer (DKFiltSoft) were more likely to receive sexual material (b = 2.584, p < .01). The other significant predictors of receipt of sexual material for males were the following: use of Facebook (b = 1.004, p < .05); having a parent in the room during Internet use (ParInRm) (b = .818, p < .05); unsure of filtering or blocking software on the computer (DKFiltSoft) (b = 2.584, p < .05); increased respect for parents (RespectParents) (b = .544, p < .01); having parents yell at you (Yell) (b = .228, p < .05); and having privileges taken away by parents (Privileges) (b = .424, p < .001).

Table 27. Logistic Regression Estimates for the Dependent Variable Receipt Sexually Explicit Material During the High School Senior Time Period

Variable	Male Model (N = 195)		Variable	Female Model (N = 288)	
	B(SE)	Exp(B)		B(SE)	Exp(B)
ChatHour	2.012(.625)	7.477***	Shop	.664(.347)	1.943
Facebook	1.044(.414)	2.758*	Social	1.482(.804)	4.400
SchLab	2.355(.922)	10.537*	Other	1.309(.472)	3.701**
ParInRm	.818(.413)	2.257*	ProvidedInfo	.163(.062)	1.777*
RestrictAdult	-1.384(.498)	.250**	SibInRm	-.488(.324)	.614
DKFiltSoft	2.584(.827)	13.256**	OthInRm	.808(.474)	2.242
RespectParents	.544(.162)	1.772**	RestrictTime	.598(.390)	1.819
Grades	.126(.093)	1.135	RestrictCMC	.799(.569)	2.222
Yell	.228(.105)	1.257*	White	-.972(.394)	.378*
Nag	-.139(.101)	.870	ShareFriends	-.122(.085)	.885
Privileges	.424(.131)	1.528***	Constant	-1.765(1.092)	.171
Constant	-8.931(1.786)	.000***			
-2 Log-likelihood	163.290		-2 Log-likelihood	262.785	
Model Chi-Square	68.713***		Model Chi-Square	34.626***	
Cox & Snell R^2	.297		Cox & Snell R^2	.113	
Nagelkerke R^2	.427		Nagelkerke R^2	.176	

* p < .05, ** p < .01, *** p < .001

Females, on the other hand, had few statistically significant predictors of the dependent variable. Female respondents who used the Internet for activities designated as "Other" (Other) were almost 4 times more likely to be victimized [Exp(B) = 3.701], and those who provided personal information to online contacts (ProvidedInfo) were about 78% more likely to receive unwanted sexually explicit material online for each type of information provided [Exp(B) = 1.777]. White females were less likely to receive sexually explicit material (b = -.972, p < .05), which was a similar finding from the analysis with the total sample, but suggests that the effect of race was stronger for females than for males.

Logistic regression estimates for males and females for the dependent variable "receipt of non-sexual harassment" during the high school senior time period are presented in Table 28. Variables retained at the .20 level were shown to explain 22.0% to 31.6% of the variation in the dependent variable for males, but only 12.5% to 17.1% for females. As can be seen in the table, males and females shared one statistically significant predictor. Both male (b = .211, p < .01) and female (b = .179, p < .01) respondents who provided personal information to online contacts (ProvidedInfo) had a greater likelihood of receipt of non-sexual harassment. Planning travel online (Travel) also increased the likelihood a male in the sample would receive non-sexual harassment (b = .786, p < .05), and having a teacher in the room during Internet use (TeachInRm) decreased the likelihood a female would receive non-sexual harassment (b = -.942, p < .05). Also, males in the sample who reported using the Internet to socialize (Social) (b = 2.240, p < .05) and those more able to share thoughts and feelings with friends (ShareFriends) (b = .228, p < .05) were shown to have an increased likelihood of victimization. Other significant predictors for males were as follows: increased use of instant messaging (IMHours) (b = .098, p < .05); increased use of social networking websites (SNWHours) (b = -.177, p < .05); and the respondent's grade point average at the time of high school graduation (b = -.406, p < .05).

Full logistic regression models for males and females for the dependent variable "receipt of sexual solicitation" during the high school senior time period are presented in Table 29. Here, variables retained at the .20 level were shown to explain 26.7% to 52.8% of the

Table 28. Logistic Regression Estimates for the Dependent Variable Receipt of Non-Sexual Harassment During the High School Senior Time Period

Variable	Male Model (N = 195)		Variable	Female Model (N = 288)	
	B(SE)	Exp(B)		B(SE)	Exp(B)
Travel	.786(.393)	2.195*	Social	1.418(.783)	4.128
Design	.850(.474)	2.339	SNWHours	.068(.037)	1.070
Social	2.240(1.120)	9.365*	ProvidedInfo	.179(.058)	1.196**
EmHours	.337(.197)	1.401	TeachInRm	-.942(.460)	.390*
IMHours	.098(.049)	1.103*	OthInRm	.807(.448)	2.241
SNWInfo	-.177(.075)	.838*	RestrictCMC	.939(.529)	2.558
ProvidedInfo	.211(.069)	1.235**	Age	.436(.720)	1.547
RestrictTime	-1.067(.609)	.344	Constant	-5.526(1.020)	.000*
GPA	-.406(.170)	.666*			
ShareParents	-.117(.090)	.890			
ShareFriends	.228(.129)	1.257*			
Constant	-4.066(1.429)	.017**			
-2 Log-likelihood	180.595		-2 Log-likelihood	337.224	
Model Chi-Square	47.845***		Model Chi-Square	38.356***	
Cox & Snell R^2	.220		Cox & Snell R^2	.125	
Nagelkerke R^2	.316		Nagelkerke R^2	.171	

* $p < .05$, ** $p < .01$, *** $p < .001$

variation in the dependent variable for males, but only 13.1% to 25.4% for females. Males and females again shared one statistically significant predictor, which was the same as in the previous models. Both male [Exp(B) = 1.400] and female [Exp(B) = 1.330] respondents who provided personal information to online contacts (ProvidedInfo) had a greater likelihood of receipt of sexual solicitation.

Males and females also had separate predictors of the dependent variable. The only other significant predictor for females was posting personal information on a social networking website (SNWInfo), which was shown to increase the likelihood of receipt of sexual solicitation (b = .271, p < .05). On the other hand, males had multiple predictors. Having restrictions on viewing adult websites (RestrictAdult) (b = -2.316, p < .01) decreased the likelihood of receipt of sexual solicitation for male respondents. As seen in other models, males who had privileges revoked by parents (Privileges) were at a greater likelihood of victimization (b = .394, p < .05); however, the issue of temporal ordering should again be considered along with the direction of the slope. The three other significant predictors of receipt of sexual solicitation for males were the following: sharing thoughts and feelings with friends (ShareFriends) (b = -.357, p < .05); having respect for parents (RespectParents) (b = .480, p < .05); and having respect for teachers (RespectTeachers) (b = -.349, p < .05). For females, providing personal information on a social networking website was positive and significant (b = .271; p < .05).

The remaining four dependent variables in the study examined the formation of various relationships with online contacts. Table 30 presents the logistic regression estimates for males and females for the dependent variable "formation of a virtual relationship with a person met online" during the high school senior time period. Variables retained at the .20 level were shown to explain a respectable 28.9% to 58.9% of the variation in the dependent variable for males, as well as 33.8% to 54.6% for females. Males and females shared two significant predictors, in a similar manner as the model examining the total sample. Both male (b = .206, p < .05) and female (b = .322, p < .01) respondents who provided personal information to online contacts (ProvidedInfo) experienced a greater likelihood of formation of a virtual relationship with an online contact, as did males (b = 1.364, p < .05) and females (b = 2.344, p < .001) who communicated with online

Table 29. Logistic Regression Estimates for the Dependent Variable Receipt Sexual Solicitation During the High School Senior Time Period

Variable	Male Model (N = 195)		Variable	Female Model (N = 288)	
	B(SE)	Exp(B)		B(SE)	Exp(B)
Gaming	-1.048(.565)	.351	Travel	-.856(.467)	.425
MySpace	1.280(.744)	3.598	Shop	.853(.474)	2.347
ProvidedInfo	.336(.101) 1	.400**	SNWInfo	.271(.115)	1.311*
OthInRm	2.277(.886)	9.747*	ProvidedInfo	.285(.072)	1.330***
RestrictAdult	-2.316(.887)	.099**	NoOneRm	.670(.418)	1.954
DKActMon	1.404(.792)	4.073	RestrictAdult	.654(.434)	1.924
GPA	-.345(.259)	.708	DKFiltSoft	.683(.510)	1.980
ShareFriends	-.357(.172)	.700*	Constant	-4.839(.797)	.008***
RespectParents	.480(.204)	1.616*			
RespectTeach	-.349(.156)	.706*			
Nag	-.289(.166)	.749			
Privileges	.394(.170)	1.482*			
Constant	-1.572(1.994)	.208			
-2 Log-likelihood	76.814		-2 Log-likelihood	168.649	
Model Chi-Square	60.611***		Model Chi-Square	40.458***	
Cox & Snell R^2	.267		Cox & Snell R^2	.131	
Nagelkerke R^2	.528		Nagelkerke R^2	.254	

* $p < .05$; ** $p < .01$; *** $p < .001$

Table 30. Logistic Regression Estimates for the Dependent Variable Formation of Virtual Relationship During the High School Senior Time Period

Variable	Male Model (N = 195)		Variable	Female Model (N = 288)	
	B(SE)	Exp(B)		B(SE)	Exp(B)
Shop	-1.306(.490)	.271**	Design	.657(.499)	1.929
Facebook	1.232(.523)	3.429*	Shop	.682(.464)	1.978
Comm	1.364(.562)	3.912*	IMHours	.171(.069)	1.186*
ProvidedInfo	.206(.081)	1.228*	SNWHours	-.139(.075)	.870
NoOneRm	.820(.460)	2.269	MySpace	1.591(.641)	4.907*
Age	.683(.473)	1.980	Comm	2.344(.546)	10.248***
ShareFriends	.323(.191)	1.381	ProvidedInfo	.322(.096)	1.380**
EnjoyFriends	-.460(.262)	.631	FriInRm	-1.149(.476)	.317*
RespectTeach	-.224(.128)	.800	SibInRm	.694(.474)	2.002
Nag	-.307(.111)	.736**	NoOneRm 1	.263(.485)	3.534**
Privileges	.211(.109)	1.235	DKFiltSoft	1.329(.611)	3.777*
Constant	-11.926(8.954)	.000	Age	-1.245(.472)	.288**
			Grades	.585(.263)	1.795*
			Succeed	-.289(.196)	.790
			Yell	-.139(.104)	.870
			Nag	.150(.105)	1.162
			Constant	14.262(8.658)	.000*

-2 Log-likelihood	63.134		-2 Log-likelihood	118.817	
Model Chi-Square	64.511***		Model Chi-Square	40.458***	
Cox & Snell R²	.289		Cox & Snell R²	.338	
Nagelkerke R²	.589		Nagelkerke R²	.546	

* p < .05, ** p < .01, *** p < .001

contacts (Comm). Participation in these behaviors appeared to have a greater impact for females, as providing personal information to online contacts (ProvidedInfo) increased the occurrence of the dependent variable by 38% [Exp(B) = 1.380] for each type of information provided, and communicating with online contacts (Comm) increased the likelihood of forming a virtual relationship by 10 times [Exp(B) = 10.248].

Males and females also had several different statistically significant predictors of formation of a virtual relationship. Specifically with regard to males in the sample, use of Facebook (b = 1.232, p < .05) increased the likelihood of forming a virtual relationship, while nagging by parents (Nag) decreased the likelihood of forming a virtual relationship (b = -.307, p < .01). For females, having no one in the room during Internet use (NoOnRm) (b = 1.263, p < .01), use of MySpace (b = 1.591, p < .05), and use of instant messaging (IMHours) (b = .171, p < .05) increased the likelihood of formation of a virtual relationship. Other significant predictors of formation of a virtual relationship with online contacts for females were as follows: having a friend in the room during Internet use (FriInRm) (b = -1.149, p < .05); being unsure if filtering or blocking software was installed on the respondent's computer (DKFiltSoft) (b = 1.329, p < .05); age (b = -1.245, p < .01); and the importance placed on grades (Grades) (b = .585, p < .05).

Table 31 presents the logistic regression estimates for males and females for the dependent variable "participation in offline contact with a person met online" during the high school senior time period. Variables retained at the .20 level were shown to explain a rather large 42.5% to 64.2% of the variation in the dependent variable for males, as well as 41.8% to 63.0% for females. Males and females again shared two positive statistically significant predictors. Both male (b = .236, p < .05) and female (b = .498, p < .001) respondents who provided personal information to online contacts (ProvidedInfo) experienced a greater likelihood of participation in offline contact. In addition, males (b = 2.769, p < .05) and females (b = 1.506, p < .001) who communicated with online contacts (Comm) experienced a greater likelihood. Respondents who were unaware if they were monitored during Internet use (DKActMon) experienced a different likelihood of participating in offline contact based on gender.

Table 31. Logistic Regression Estimates for the Dependent Variable Participation in Offline Contact During the High School Senior Time Period

Variable	Male Model (N = 195) B(SE)	Exp(B)	Variable	Female Model (N = 288) B(SE)	Exp(B)
SNWHours	.226(.100)	1.253*	MySpace	1.258(.761)	3.519
MySpace	2.003(.748)	7.410**	NoPrivSNW	-1.763(.619)	.174**
Comm	2.769(.704)	15.940***	SNWInfo	-.279(.153)	.757
ProvidedInfo	.236(.101)	1.266*	Comm	1.506(.551)	4.507**
ParInRm	-1.337(.654)	.263*	ProvidedInfo	.498(.103)	1.646***
FriInRm	1.023(.675)	2.782	FriInRm	-1.356(.462)	.258**
TeachInRm	1.257(.847)	3.517	OthInRm	-3.271(1.376)	.038*
RestrictTime	1.063(.767)	2.896	RestrictAdult	-1.251(.577)	.286*
RestrictAdult	-1.199(.670)	.301	RestrictCMC	2.161(.867)	8.861*
DKActMon	-2.506(.952)	.082**	DKActMon	1.084(.531)	2.958*
ShareFriends	.479(.267)	1.615	GPA	.396(.290)	1.485
EnjoyFriends	-.488(.310)	.614	ShareParents	.143(.092)	1.154
RespectTeach	.299(.176)	1.348	RespectParents	-.192(.145)	.825
Participate	-.153(.117)	.858	RespectTeach	-.212(.146)	.809
Nag	-.163(.121)	.849	Participate	.862(.329)	2.367**
Privileges	.361(.133)	1.434**	Grades	.359(.193)	1.432
Constant	-3.242(1.875)	.039	Constant	-2.973(1.508)	.051*
-2 Log-likelihood	102.689		-2 Log-likelihood	105.315	
Model Chi-Square	106.935***		Model Chi-Square	104.319***	
Cox & Snell R²	.425		Cox & Snell R²	.418	
Nagelkerke R²	.642		Nagelkerke R²	.630	

* p < .05, ** p < .01, *** p < .001

163

Males who were unaware were less likely (b = -2.506, p < .01) and females who were unaware were more likely (b = 1.085, p < .01) to participate in offline contact.

Respondents also had various unique statistically significant predictors of participation in offline contact based on gender. For males in the sample, use of MySpace (b = 2.003, p < .01) and greater hours per week use of social networking websites (SNWHours) (b = .226, p < .05) increased the likelihood of participation in offline contact. Other significant predictors of participating in offline contact for males were the following: having a parent in the room during Internet use (ParInRm) (b = -1.337, p < .05) and having privileges revoked by parents (Privileges) (b = .361, p < .01). For females, respondents who used a non-privatized social networking website (NoPrivSNW) were less likely to participate in offline contact with a person met online (b = -1.763, p < .01). Moreover, females who had a friend in the room (FriInRm) (b = -1.356, p < .01) or a person designated as "Other" (Other) (b = -3.271, p < .05) also were less likely to participate in off-line contact, as were those who had restrictions on viewing adult websites (RestrictAdult) (b = -1.251, p < .05). The remaining positive and significant predictors of the dependent variable for females were the following: restrictions on the use of CMCs (b = 2.161, p < .05) and an importance placed on participation in school activities (b = .862, p < .01).

Logistic regression estimates for males and females for the dependent variable "face-to-face contact with an online contact" during the high school senior time period are presented in Table 32. Variables retained at the .20 level were shown to explain 34.0% to 59.0% of the variation in the dependent variable for males, as well as 26.9% to 46.7% for females. Males and females again shared the same two positive significant predictors. Both male (b = .528, p < .001) and female (b = .291, p < .01) respondents who provided personal information to online contacts (ProvidedInfo) had a greater likelihood of participation in face-to-face offline contact, as did males (b = 1.931, p < .05) and females (b = 1.671, p < .05) who communicated with online contacts (Comm).

Table 32. Logistic Regression Estimates for the Dependent Variable Face-to-Face Contact During the High School Senior Time Period

Variable	Male Model (N = 195) B(SE)	Exp(B)	Variable	Female Model (N = 288) B(SE)	Exp(B)
Other	1.525(.977)	4.597	Comm	1.671(.656)	5.320*
Facebook	1.602(.813)	4.961*	ProvidedInfo	.291(.090)	1.338**
Comm	1.931(.966)	6.895*	FriInRm	-1.941(.652)	.144**
ProvidedInfo	.528(.143)	1.696***	TeachInRm	1.785(.964)	5.960
ParInRm	-1.172(.787)	.310	RestrictAdult	-2.999(.898)	.050**
FriInRm	1.683(.765)	5.381	White	2.288(1.188)	9.857
ActMon	1.753(.989)	5.770	Grades	.726(.278)	2.088**
FiltSoft	-1.210(.782)	.298	Succeed	-.345(.193)	.708
White	-2.946(.988)	.053**	Yell	.119(.090)	1.127
GPA	-.455(.360)	.634	Constant	-2.100(1.181)	.122
ShareParents	.339(.184)	1.403			
EnjoyFriends	-.453(.316)	.636			
RespectParents	-.667(.288)	.513*			
Grades	-.803(.319)	.448*			
Succeed	.634(.252)	1.886*			
Yell	-.286(.156)	.752			
Constant	3.439(2.395)	1.150			
-2 Log-likelihood	86.477		-2 Log-likelihood	106.249	
Model Chi-Square	80.595***		Model Chi-Square	61.686***	
Cox & Snell R²	.340		Cox & Snell R²	.269	
Nagelkerke R²	.590		Nagelkerke R²	.467	

* p < .05, ** p < .01, *** p < .001

Males and females also had different predictors of participation in face-to-face contact. Males who used Facebook (b = 1.602, p < .05) and those who reported being more interested in succeeding in school (Succeed) (b = .634, p < .05) were more likely to participate in face-to-face contact with a person met online. Alternatively, white (White) males were less likely to participate in face-to-face contact with a person met online (b = -2.946, p < .01). Two other variables also were negative significant predictors of participating in face-to-face contact with a person met online for males: respect for parents (RespectParents) (b = -.667, p < .05) and placing an importance on grades (Grades) (b = -.803, p < .05). As for the female respondents, those who felt grades were important (Grades) (b = .726, p < .01) experienced an increased likelihood of participating in the behavior. Females who had a friend in the room (FriInRm) (b = -1.941, p < .01) or restrictions on the viewing of adult websites (RestrictAdult) (b = -2.999, p < .01) were significantly less likely to participate in offline contact.

The last models examined for the high school senior time period, presented in Table 33, include the logistic regression estimates for males and females for the dependent variable "sexual encounter with an online contact." Variables retained at the .20 level were shown to explain 26.4% to 52.2% of the variation in the dependent variable for males, but only 16.3% to 31.7% for females. Males and females had two shared statistically significant predictors. Both males (b = .306, p < .01) and females (b = .328, p < .001) who provided one or more types of personal information to people met online were more likely to participate in a sexual encounter with a person met online. Conversely, males (b = -.367, p < .05) and females (b = -.258, p < .05) who felt comfortable sharing thoughts and feelings with friends (ShareFriends) were less likely to participate in a sexual encounter with a person online.

Males and females also had other unique statistically significant predictors of the dependent variable. With regard to males, respondents who had a person designated as "Other" in the room during Internet use (OthInRm) experienced an increased likelihood of a sexual encounter with an online contact (b = 2.374, p < .01), while those who had restrictions on viewing adult websites (RestrictAdult) were less likely

Table 33. Logistic Regression Estimates for the Dependent Variable Sexual Encounter During the High School Senior Time Period

Variable	Male Model (N = 195)		Variable	Female Model (N = 288)	
B(SE)	Exp(B)		B(SE)	Exp(B)	
MySpace	1.463(.761)	4.319	Travel	-.883(.492)	.414
ProvidedInfo	.306(.115)	1.357**	Shop	.760(.485)	2.138
OthPI	2.194(1.403)	8.970	SNWInfo	.289(.118)	1.335*
OthInRm	2.374(.898)	10.737**	ProvidedInfo	.328(.076)	1.385***
RestrictAdult	-2.882(1.010)	.056**	RestrictAdult	.741(.453)	2.097
ActMon	1.533(1.138)	4.630	ActMon	-.923(.692)	.897
DKActMon	1.791(.819)	5.996	GPA	.536(.236)	1.710*
GPA	-.329(.260)	.709	ShareFriends	-.258(.113)	.772*
ShareFriends	-.367(.166)	.693*	Participate	.118(.094)	1.125
RespectParents	.445(.236)	1.560	Grades	.397(.196)	1.488*
RespectTeach	-.321(.153)	.725*	Constant	-7.292(1.963)	.001***
Nag	-.310(.167)	.734			
Privileges	.367(.170)	1.476*			
Constant	-2.164(1.895)	.115			
-2 Log-likelihood	77.267		-2 Log-likelihood	157.703	
Model Chi-Square	59.798**		Model Chi-Square	51.404***	
Cox & Snell R^2	.264		Cox & Snell R^2	.163	
Nagelkerke R^2	.522		Nagelkerke R^2	.317	

* $p < .05$; ** $p < .01$; *** $p < .001$

167

to participate in a sexual encounter (b = -2.882, p < .01). Having respect for teachers (RespectTeachers) (b = -.321, p < .05) and having privileges revoked by parents (Privileges) (b = .367, p < .05) were also significant predictors of the dependent variable for males. Females, on the other hand, who posted personal information on their social networking websites (SNWInfo) (b = .289, p < .05) and had higher grade point averages at the time of high school graduation (GPA) (b = .536, p < .05) were more likely to participate in a sexual encounter. Placing an importance on grades (Grades) (b = .397, p < .05) also was shown to be a positive and significant predictor of a sexual encounter with an online contact for females.

College Freshman Time Period

Table 34 presents the logistic regression estimates for the dependent variable "receipt of unwanted sexually explicit material" during the college freshman time period. The first block of the model included the first set of independent variables, representing the theoretical construct of exposure to motivated offenders (Research, Gaming, Travel, Design, Shop, Social, Other, EmHours, IMHours, ChatHour, SNWHours, MySpace, and Facebook). Variables retained at the .20 level were shown to explain only 4.9% to 10.1% of the variation in the dependent variable. Four variables emerged as significant predictors (p < .05). Respondents who performed research online (Research) (b = -1.828, p < .05) and those who used Facebook as a social networking website (b = -1.017, p < .05) were less likely to receive unwanted sexually explicit material. Conversely, respondents who performed website design (Design) (b = 1.004, p < .01) and those who used chat rooms (ChatHour) were more likely to receive unwanted sexually explicit material (b = 1.358, p < .05). Users of chat rooms were almost 4 times more likely to be victimized [Exp(B) = 3.887].

The second set of independent variables, representing the theoretical construct of target suitability (NoPriv, SNWInfo, Comm, and ProvidedInfo), added another to the list of significant predictors of receipt of unwanted sexually explicit material. In order to deal with multicollinearity among target suitability variables representing the types of information that can be posted on a social networking website and provided to online contacts, the variables "SNWInfo" (α = .662) and "ProvidedInfo" (α = .913) were created as additive measures of the

Table 34. Logistic Regression Estimates for the Dependent Variable Receipt of Unwanted Sexually Explicit Material During the College Freshman Time Period (N = 483)

Variable	Exposure to Motivated Offenders		Target Suitability		Lack of Capable Guardianship		Control Variables	
	B(SE)	Exp(B)	B(SE)	Exp(B)	B(SE)	Exp(B)	B(SE)	Exp(B)
IntHours	.018(.011)	1.018	.018(.011)	1.018	.021(.011)	1.021	.020(.011)	1.021
Research	-1.828(.756)	.161*	-1.712(.760)	.181*	-1.666(.896)	.189	-1.300(.922)	.273
Design	1.004(.370)	2.730**	.897(.378)	2.451*	.797(.401)	2.220*	.674(.405)	1.963
Social	1.006(.785)	2.734	—	—	—	—	—	—
ChatHour	1.358(.561)	3.887*	1.348(.563)	3.849*	1.376(.581)	3.961*	1.525(.594)	4.597*
Facebook	-1.017(.436)	.362*	-1.398(.487)	.247**	-1.054(.549)	.349	-1.215(.548)	.297*
SNWInfo			.189(.090)	1.208*	.176(.096)	1.192	.176(.097)	1.192
FriInRm					-.784(.357)	.457*	-.708(.358)	.493*
OthInRm					1.553(.343)	4.727***	1.531(.345)	4.623***
ActMon					1.167(.795)	3.214	—	—
Succeed							.353(.152)	1.423*
Yell							.096(.063)	1.101
Constant	-1.191(1.028)	.304	-1.030(.892)	.357	-1.241(1.043)	.289	-4.719(1.786)	.009**
-2 Log-likelihood	296.841		294.401		269.590		263.465	
Model Chi-Square	24.156***		26.596**		51.407***		57.532	
Cox & Snell R^2	.049		.054		.101		.113	
Nagelkerke R^2	.101		.110		.208		.232	

* $p < .05$; ** $p < .01$; *** $p < .001$

169

original dichotomous variables. This model was shown to explain a modest range of 5.4% to 11.0% of the variation in the dependent variable.After the addition of target suitability variables, respondents who performed research online (Research) (b = -1.712, p < .05) and those who used Facebook as a social networking website (b = -1.398, p < .01) continued to be significantly less likely to receive unwanted sexually explicit material. Survey respondents who participated in website design (Design) (b = .897, p < .05) and those who used chat rooms (ChatHour) (b = 1.348, p < .05) continued to be more likely to receive the material. Furthermore, the model indicated that freshmen who provided personal information on their social networking websites (SNWInfo) were also more likely to be victimized in this manner (b = .189, p < .05).

Addition of the third set of independent variables, representing the theoretical construct of lack of capable guardianship (LivRm, Dorm, ParInRm, FriInRm, SibInRm, OthInRm, NoOneRm, RestrictAdult, RestrictOther, ActMon, DKActMon, FiltSoft, DKFiltSoft) produced two additional statistically significant variables (p < .05). Variables retained at the .20 level were shown to explain 10.1% to 20.8% of the variation in the dependent variable. While website design (Design) (b = .797, p < .05) and use of chat rooms (ChatHour) (b = 1.376, p < .05) remained as significant positive predictors, researching online (Research), use of Facebook, and providing personal information on a social networking website (SNWInfo) were no longer significant in the model. However, having a friend in the room during Internet use (FriInRm) decreased the likelihood of receiving unwanted sexually explicit material (b = -.784, p < .05), while having a person designated as "Other" in the room during Internet use (OthInRm) increased the likelihood by more than 4 times [Exp(B) = 4.727].

Control variables (Sex, Age, White, LivDorm, LivOther, GPA, ShareParents, ShareFriends, EnjoyFriends, RespectParents, RespectTeachers, Participate, Grades, Succeed, Yell, Nag, and Privileges) were added to create the full logistic regression model presented in the final column of Table 34. The full model was shown to explain 11.3% to 23.2% of the variation in the dependent variable. Use of chat rooms (ChatHour) (b = 1.525, p < .05) and having a person designated as "Other" in the room during Internet use (OthInRm) (b = 1.531, p < .05) continued to increase the likelihood of receipt of unwanted sexually explicit material. Use of Facebook (b = -1.215, p < .05) and having a friend in the room during Internet use (FriInRm) (b = -.708, p < .05) decreased the likelihood of receiving the material. Additionally, a stronger desire to succeed in school (Succeed) was associated with an increased likelihood of receiving unwanted sexually explicit material online (b = .353, p < .05).

Logistic regression estimates for the dependent variable "receipt of non-sexual harassment" during the college freshman time period are presented in Table 35. After the insertion of the first set of independent variables, representing the theoretical construct of exposure to motivated offenders, only one variable emerged as significant predictor: planning travel on the Internet. Variables retained at the .20 level explained a rather low 3.3% to 6.0% of the variation in the dependent variable. Respondents who planned travel on the Internet (Travel) were more likely to receive non-sexual harassment online (b = .625, p < .05).

After implementation of the second set of independent variables, representing the theoretical construct of target suitability, again only one variable displayed statistical significance in the model. Variables retained at the .20 level were shown to explain only 4.2% to 7.7% of the variation in the dependent variable. Communicating with people online (Comm) was now shown to significantly increase the likelihood of receiving non-sexual harassment (b = 6.04, p < .05).

Only two variables emerged as significant predictors after the addition of the third set of independent variables, representing the theoretical construct of lack of capable guardianship. Variables retained at the .20 level again were shown to explain a low 4.8% to 8.8% of the variation in the dependent variable. Communicating with people online (Comm) continued to increase the likelihood of receipt of non-sexual harassment (b = .640, p < .05); respondents who communicated with people online were approximately 90% more likely to be victimized in this manner [Exp(B) = 1.896]. Moreover, spending greater hours per week using social networking websites (SNWHours) also significantly increased the likelihood of receiving non-sexual harassment online (b = .062, p < .05).

Table 35. Logistic Regression Estimates for the Dependent Variable Receipt of Non-Sexual Harassment During the College Freshman Time Period (N = 483)

Variable	Exposure to Motivated Offenders		Target Suitability		Lack of Capable Guardianship		Control Variables	
	B(SE)	Exp(B)	B(SE)	Exp(B)	B(SE)	Exp(B)	B(SE)	Exp(B)
Travel	.625(.278)	1.868*	.521(.281)	1.688	.541(.286)	1.718	.717(.292)	2.048*
Social	.990(.768)	2.692	.960(.775)	2.611	1.018(.783)	2.767	1.113(.811)	3.044
SNWHours	.050(.028)	1.051	.042(.029)	1.043	.062(.029)	1.064*	—	—
MySpace	.417(.294)	1.517	—	—	—	—	—	—
Facebook	-.637(.446)	.529	-.886(.476)	.412	-.744(.470)	.475	-.893(.477)	.409
SNWInfo			.116(.087)	1.123	—	—	—	—
Comm			.604(.281)	1.829*	.640(.286)	1.896*	.882(.294)	2.415**
RestrictAdult					-1.051(.659)	.350	-1.221(.679)	.295
Sex							1.166(.340)	3.209**
ShareParents							.136(.071)	1.146
ShareFriends							-.266(.100)	.766**
EnjoyFriends							.382(.145)	1.465**
Nag							.140(.056)	1.150*
Privileges							-.125(.097)	.882
Constant	-.3046(.780)	.048***	-3.390(.807)	.034***	-3.114(.786)	.044***	-6.070(1.410)	.002***
-2 Log-likelihood	365.332		360.417		357.465		336.489	
Model Chi-Square	15.942**		20.856**		23.089**		44.785	
Cox & Snell R²	.033		.042		.048		.089	
Nagelkerke R²	.060		.077		.088		.162	

* p < .05; ** p < .01; *** p < .001

Addition of the control variables created the full logistic regression model presented in Table 35. Variables retained at the .20 level were shown to explain 8.9% to 16.2% of the variation in the dependent variable. Multiple variables were revealed as significant predictors, particularly those designated as control variables. Planning travel online (Travel) reemerged as a significant positive predictor of victimization (b = .717, p < .05), and communicating with people online (Comm) remained significant (b = .882, p < .01). With regard to gender, females were shown to be over 3 times more likely to receive non-sexual harassment [Exp(B) = 3.209]. Sharing thoughts and feelings with friends (ShareFriends) decreased the likelihood of non-sexual harassment (b = -.266, p < .01), while enjoying spending time with friends (EnjoyFriends) increased the likelihood of this type of victimization (b = .382, p < .01). Finally, being nagged more often by parents (Nag) also increased the likelihood of receipt of non-sexual harassment (b = .140, p < .05).

Table 36 presents the logistic regression estimates for the dependent variable "receipt of sexual solicitation" during the college freshman time period. Only one variable was statistically significant after the insertion of independent variables representing exposure to motivated offenders. Variables retained at the .20 level also were shown to explain only 1.9% to 5.4% of the variation in the dependent variable. Respondents who participated in website design on the Internet (Design) were more likely to receive sexual solicitation (b = .985, p < .05). Furthermore, the addition of the second set of independent variables, representing the theoretical construct of target suitability, resulted in no independent variables being statistically significant.

Three variables emerged as significant predictors after the addition of the third set of independent variables, representing the theoretical construct of lack of capable guardianship. Variables retained at the .20 level were shown to explain 5.4% to 15.1% of the variation in the dependent variable. Website design (Design) reemerged as a statistically significant positive predictor of receipt of sexual solicitation (b = .923, p < .05). Communicating with people online (Comm) (b = .920, p < .05), as well as having a person designated as "Other" in the room during Internet use (OthInRm) (b = .927, p < .05),

Table 36. Logistic Regression Estimates for the Dependent Variable Receipt of Sexual Solicitation During the College Freshman Time Period (N = 483)

Variable	Exposure to Motivated Offenders		Target Suitability		Lack of Capable Guardianship		Control Variables	
	B(SE)	Exp(B)	B(SE)	Exp(B)	B(SE)	Exp(B)	B(SE)	Exp(B)
Design	.985(.451)	2.677*	.840(.457)	2.317	.923(.470)	2.518*	.896(.481)	2.450
EmHours	.180(.133)	1.197	.175(.133)	1.191	—	—	—	—
ChatHour	1.262(.674)	3.462	1.037(.684)	2.822	1.177(.709)	3.245	1.194(.724)	3.299
Comm			.773(.424)	2.167	.920(.447)	2.508*	.922(.456)	2.514*
ParInRm					1.487(.789)	4.423	1.927(.869)	6.870*
SibInRm					-1.846(1.214)	.158	-2.496(1.372)	.084
OthInRm					.927(.452)	2.526*	.957(.456)	2.605*
DKActMon					.955(.508)	2.598	.806(.528)	2.240
DKFiltSoft					-1.380(.834)	.252	-1.560(.856)	.210
LivDorm							1.092(.811)	2.980
Nag							-.133(.102)	.875
Privileges							.289(.123)	1.335*
Constant	-3.554(.446)	.029***	-3.936(.514)	.020***	-3.911(.432)	.020***	-.4873(.908)	.008***
-2 Log-likelihood	204.280		200.836		186.840		181.650	
Model Chi-Square	9.422*		12.867*		26.862**		32.052**	
Cox & Snell R²	.019		.026		.054		.064	
Nagelkerke R²	.054		.074		.151		.180	

* p < .05; ** p < .01; *** p < .001

174

were also positive predictors of victimization. All three of these noted independent variables indicated that respondents who participated in Internet use in these manners were more than 2 times more likely to be victimized [Exp(B) = 2.518; Exp(B) = 2.508; Exp(B) = 2.526, respectively].

Finally, control variables were added to create the full logistic regression model presented in Table 47. Variables retained at the .20 level were found to explain 6.4% to 18.0% of the variation in the dependent variable. Multiple variables emerged as significant predictors of receipt of sexual solicitation. Both communicating with others online (Comm) (b = .922, p < .05) and having a person designated as "Other" in the room during Internet use (OthInRm) (b = .957, p < .05) remained significant positive predictors. Respondents who had parents in the room during Internet use (ParInRm) (b = 1.927, p < .05), as well as those who more often had privileges taken away by parents (Privileges) (b = .289, p < .05), were also more likely to receive sexual solicitation. As previously discussed in other models, it could be that temporal ordering is an issue in this analysis. It is possible that monitoring by parents, as well as revoked privileges, could be a result of previous Internet behaviors and experiences.

The remaining dependent variables examined various relationships formed with people met online. Table 37 presents the logistic regression estimates for the dependent variable "formation of a virtual relationship" with an online contact during the college freshman time period. Only one variable was significant after the insertion of the first set of independent variables, representing the theoretical construct of exposure to motivated offenders: spending time in chat rooms. Variables retained at the .20 level were shown to explain 6.9% to 11.1% of the variation in the dependent variable. Respondents who spent time in chat rooms (ChatHour) were over four times more likely to form virtual relationships with online contacts [Exp(B) = 4.436].

After the second set of independent variables, representing the theoretical construct of target suitability, were added to the model, chat room use was no longer significant. Variables retained at the .20 level were shown to explain 19.8% to 31.7% of the variation in the dependent variable. Three target suitability variables did emerge as statistically significant predictors of the dependent variable. The first statistically significant positive predictor was the use of a non-privatized social networking website (NoPriv) (b = .574, p < .05).

Table 37. Logistic Regression Estimates for Formation of Virtual Relationship During the College Freshman Time Period (N = 483)

Variable	Exposure to Motivated Offenders		Target Suitability		Lack of Capable Guardianship		Control Variables	
	B(SE)	Exp(B)	B(SE)	Exp(B)	B(SE)	Exp(B)	B(SE)	Exp(B)
IntHours	.018(.009)	1.018	.017(.010)	1.017	.021(.010)	1.021*	.028(.010)	1.028**
ChatHour	1.490(.492)	4.436**	1.233(.544)	3.399	1.482(.590)	4.402*	1.009(.607)	2.742
SNWHours	.052(.026)	1.053	.054(.029)	1.056	.049(.029)	1.050		
MySpace	.468(.265)	1.597	—	—	—	—	—	—
NoPriv			.574(.272)	1.755*	.689(.280)	1.991*	.711(.290)	2.036*
Comm			1.279(.314)	3.593***	1.327(.321)	3.769***	1.440(.335)	4.221***
ProvidedInfo			.259(.054)	1.296***	.270(.056)	1.311***	.306(.059)	1.358***
Dorm					1.092(.516)	2.981*	1.012(.528)	2.752
Sex							.700(.314)	2.013*
White							-.912(.358)	.402*
ShareParents							.105(.079)	1.111
ShareFriends							-.167(.086)	.846
RespectFriends							-.146(.098)	.864
Grades							.317(.179)	1.373
Succeed							-.401(.149)	.669**
Yell							-.096(.069)	.909
Constant	-2.646(.313)	.071***	-3.627(.411)	.027***	-5.074(.690)	.006***	-1.998(1.357)	.136
-2 Log-likelihood	435.402		363.860		354.338		336.367	
Model Chi-Square	34.533***		106.075***		115.596***		132.267***	
Cox & Snell R²	.069		.198		.213		.240	
Nagelkerke R²	.111		.317		.342		.385	

* p < .05, ** p < .01***, p < .001

176

In addition, respondents who communicated with people online (Comm) were more than 3 times more likely to participate in a virtual relationship [Exp(B) = 3.593], and respondents who provided personal information to online contacts (ProvidedInfo) were 29% more likely to participate in a virtual relationship for each type of information provided [Exp(B) = 1.296].

There were several new predictors of the dependent variable after the addition of the third set of independent variables, representing the theoretical construct of lack of capable guardianship. Variables retained at the .20 level were shown to explain 21.3% to 34.2% of the variation in the dependent variable. Spending more time on the Internet (IntHours) increased the likelihood of forming a virtual relationship (b = .021, p < .05), along with the use of chat rooms (ChatHour) (b = 1.482, p < .05). Use of a non-privatized social networking website (NoPriv) (b = .689, p < .05), as well as providing personal information to online contacts (ProvidedInfo) (b = .270, p < .001) continued to be positive predictors of formation of a virtual relationship. Additionally, communicating with others online (Comm) continued to be the most significant predictor of the dependent variable, as respondents who did so were almost 4 times more likely to form a virtual relationship with an online contact [Exp(B) = 3.769]. Finally, after the addition of the newest set of variables, main use of a computer in a dorm room (Dorm) also was shown to increase the likelihood of the dependent variable occurring (b = 1.092, p < .05).

Last, control variables were added to create the full logistic regression model presented in Table 37. Variables retained at the .20 level were shown to explain a respectable 24.0% to 38.5% of the variation in the dependent variable. Greater use of the Internet (IntHours) (b = .028, p < .01) and use of a non-privatized social networking website (NoPriv) (b = .711, p < .05) continued to increase the likelihood of forming a virtual relationship. As in a number of previous models, communicating with others online (Comm) (b = 1.440, p < .001) and providing personal information to online contacts (ProvidedInfo) (b = .306, p < .001) were the most statistically significant predictors of the dependent variable. With regard to the control variables, white respondents (White) (b = -.912, p < .05) and those who believed it was important to succeed in school (Succeed) (b = -.401, p < .01) were less likely to form virtual relationships with online contacts. On the other hand, female respondents (Sex) were 2 times more likely than males to form virtual relationships with online contacts [Exp(B) = 2.013].

Table 38 presents the logistic regression estimates of "participation in offline contact with a person met online" during the college freshman time period. After the insertion of the first set of independent variables, representing the theoretical construct of exposure to motivated offenders, only one variable emerged as a significant predictor. Variables retained at the .20 level were shown to explain 6.3% to 10.7% of the variation in the dependent variable. Respondents who used MySpace were significantly more likely to participate in offline contact with a person met online (b = .737, p < .05).

After the second set of independent variables (representing the theoretical construct of target suitability) were implemented in the model, use of MySpace was no longer significant. However, variables retained at the .20 level were shown to explain 20.9% to 35.5% of the variation in the dependent variable. Greater hours per week spent on social networking websites (SNWHours) was shown to increase the likelihood of the dependent variable occurring (b = .060, p < .05). Not surprisingly, as seen in past models, communicating with people online (Comm) (b = 1.466, p < .001) and providing personal information to online contacts (ProvidedInfo) (b = .298, p < .001) appeared as highly significant predictors of participating in offline contact in this partial model.

Only one new variable emerged as a significant predictor of the dependent variable after the addition of the third set of independent variables, representing the theoretical construct of lack of capable guardianship. Variables retained at the .20 level were shown to explain 21.8% to 36.9% of the variation in this model. Having a person in the room designated as "Other" (OthInRm) increased the likelihood of participating in offline contact by more than 2 times [Exp(B) = 2.273]. Greater hours per week use of social networking websites (SNWHours) (b = .062, p < .05) and providing information to online contacts (ProvidedInfo) (b = .287, p < .001) continued to increase the likelihood of participating in offline contact. Furthermore, respondents who communicated with others online (Comm) still were over 4 times more likely to participate in offline contact with a person met online [Exp(B) = 4.605].

Table 38. Logistic Regression Estimates for the Dependent Variable of Participation in Offline Contact During the College Freshman Time Period (N = 483)

Variable	Exposure to Motivated Offenders		Target Suitability		Lack of Capable Guardianship		Control Variables	
	B(SE)	Exp(B)	B(SE)	Exp(B)	B(SE)	Exp(B)	B(SE)	Exp(B)
IntHours	.016(.010)	1.016	.018(.011)	1.018	.019(.011)	1.020	.023(.011)	1.024*
IMHours	.043(.029)	1.044	—	—	—	—	—	—
SNWHours	.048(.030)	1.049	.060(.030)	1.062*	.062(.031)	1.064*	.053(.032)	1.055
MySpace	.737(.288)	2.089*	.513(.325)	1.671	.470(.327)	1.600	.467(.337)	1.595
Comm			1.466(.363)	4.330***	1.527(.367)	4.605***	1.565(.379)	4.781***
ProvidedInfo			.298(.056)	1.348***	.287(.056)	1.332***	.288(.057)	1.334***
OthInRm					.821(.353)	2.273*	.945(.368)	2.574*
White							-.613(.390)	.542
GPA							-.207(.148)	.813
Yell							-.174(.097)	.840
Nag							.170(.068)	1.186*
Privileges							.166(.098)	1.180
Constant	-3.092(.342)	.045***	-4.206(.463)	.015***	-4.438(.487)	.012***	-4.122(.690)	.016***
-2 Log-likelihood	398.439		316.752		311.534		300.024	
Model Chi-Square	31.583***		113.270***		118.488***		129.998***	
Cox & Snell R^2	.063		.209		.218		.236	
Nagelkerke R^2	.107		.355		.369		.401	

* $p < .05$, ** $p < .01$***, $p < .001$

179

The full logistic regression model, after insertion of the control variables, is presented in Table 38. Five variables were revealed as significant influences on participation in offline contact. Increased hours per week use of the Internet (IntHours) (b = .023, p < .01) and having a person designated as "Other" in the room during Internet use (OthInRm) (b = .945, p < .05) remained as positive influences on the likelihood of forming a virtual relationship. Communicating with others online (Comm) (b = 1.565, p < .001) and providing personal information to online contacts (ProvidedInfo) (b = .288, p < .001) continued to be the most significant predictors of the dependent variable. In addition, respondents who reported parents nagged them more often (Nag) were significantly more likely to participate in offline contact with a person met online (b = .170, p < .50).

Table 39 presents the logistic regression estimates for the dependent variable "participation in face-to-face contact with a person met online" during the college freshman time period. After the insertion of the first set of independent variables, representing the theoretical construct of exposure to motivated offenders, two variables emerged as significant predictors. Variables retained at the .20 level were found to explain 4.6% to 9.0% of the variation in the dependent variable. Greater use of the Internet (IntHours) increased the likelihood of participating in a face-to-face contact with a person met online (b = .031, p < .01). Moreover, respondents who used chat rooms (ChatHour) were 3 times more likely to participate in face-to-face contact with a person met online [Exp(B) = 3.027].

Predictors of the dependent variable changed after the second set of independent variables, representing the theoretical construct of target suitability, was implemented in the model. Variables retained at the .20 level now were shown to explain 13.9% to 27.3% of the variation in the dependent variable. Greater hours per week use of the Internet (IntHours) remained as a statistically significant positive predictor of participating in face-to-face contact (b = .030, p < .05), but use of a chat room (ChatHour) was no longer significant. Not surprisingly, as was seen in past models, communicating with people online (Comm) (b = 1.540, p < .001) and providing personal information to online contacts (ProvidedInfo) (b = .237, p < .001) were the most statistically significant predictors of participating in face-to-face contact in this partial model.

Table 39. Logistic Regression Estimates for the Dependent Variable Participation in Face to Face Contact During the College Freshman Time Period (N = 483)

Variable	Exposure to Motivated Offenders		Target Suitability		Lack of Capable Guardianship		Control Variables	
	B(SE)	Exp(B)	B(SE)	Exp(B)	B(SE)	Exp(B)	B(SE)	Exp(B)
IntHours	.031(.010)	1.032**	.030(.010)	1.031**	.031(.012)	1.032**	.038(.012)	1.039**
Design	.611(.365)	1.842	—	—	—	—	—	—
ChatHour	1.108(.560)	3.027*	—	—	—	—	—	—
MySpace	.550(.330)	1.733	—	—	—	—	—	—
Comm			1.540(.428)	4.665***	1.575(.434)	4.830***	1.801(.463)	6.053***
ProvidedInfo			.237(.054)	1.267***	.204(.056)	1.227***	.238(.060)	1.269***
OthInRm					.898(.387)	2.454*	1.022(.420)	2.777*
RestrictAdult					-.924(.779)	.397	—	—
RestrictOther					.981(.717)	2.667		
DKFiltSoft					-1.191(.748)	.304	-1.500(.788)	.223
White							-.769(.457)	.464
ShareFriends							-.153(.096)	.858
RespectParents							-.203(.103)	.816*
Nag							.252(.077)	1.287**
Privileges							.153(.117)	1.165
Constant	-3.430(.388)	.032***	-4.285(.493)	.014***	-4.356(.522)	.013***	-3.460(1.418)	.031*
-2 Log-likelihood	319.829		270.443		259.529		235.229	
Model Chi-Square	22.561***		72.037***		82.951***		107.251***	
Cox & Snell R²	.046		.139		.158		.199	
Nagelkerke R²	.090		.273		.311		.392	

* p < .05, ** p < .01, *** p < .001

Finally, control variables were added to create the full logistic regression model presented in Table 39. Variables retained at the .20 level were shown to explain 19.9% to 39.2 % of the variation in the dependent variable. Greater hours per week use of the Internet (IntHours) (b = .038, p < .01) and having a person designated as "Other" in the room during Internet use (OthInRm) (b = 1.022, p < .05) remained as variables that increased the likelihood of the dependent variable. Again, communicating with others online (Comm) (b = 1.801, p < .001) and providing personal information to online contacts (ProvidedInfo) (b = .238, p < .001) continued to be the most statistically significant predictors. In addition to these variables, respondents who reported parents nagged them more often (Nag) were significantly more likely to participate in face-to-face offline contact with a person met online (b = .252, p < .01). Alternatively, respondents who reported greater respect for their parents (RespectParents) were less likely to participate in the behavior (b = -.203, p < .05).

The final dependent variable examined was "participation in a sexual encounter." Table 40 presents the logistic regression estimates for this variable during the college freshman time period. After the insertion of the first set of independent variables, representing the theoretical construct of exposure to motivated offenders, only one variable emerged as a significant predictor. Variables retained at the .20 level also were shown to explain only 1.8% to 8.0% of the variation in the dependent variable. Greater use of instant messaging (IMHours) increased the likelihood of the dependent variable occurring (b = .108, p < .05).

After the second set of independent variables, representing the theoretical construct of target suitability, was implemented in the model, those retained at the .20 level were shown to explain 3.8% to 17.5% of the variation in the dependent variable. The only variable shown to be statistically significant was providing personal information to an online contact (ProvidedInfo). Respondents who provided one or more types of personal information to online contacts were 34% more likely to have a sexual encounter with an online contact for each type of information provided [Exp(B) = 1.339]. A second variable was added as a significant predictor after the addition of the third set of independent variables, representing the theoretical construct of lack of capable guardianship.

Table 40. Logistic Regression Estimates for the Dependent Variable Participation in a Sexual Encounter During the College Freshman Time Period (N = 483)

Variable	Exposure to Motivated Offenders		Target Suitability		Lack of Capable Guardianship		Control Variables	
	B(SE)	Exp(B)	B(SE)	Exp(B)	B(SE)	Exp(B)	B(SE)	Exp(B)
Gaming	.767(.598)	2.153	—	—	—	—	—	—
Other	1.237(.640)	3.446 (.053)	1.094(.656)	2.987	1.170(.661)	3.222	1.249(.734)	3.486
IMHours	.108(.052)	1.114*	.078(.057)	1.081	.102(.059)	1.108	.152(.065)	1.164*
ProvidedInfo			.292(.079)	1.339***	.283(.081)	1.327***	.410(.102)	1.506***
OthInRm					1.310(.638)	3.707*	1.225(.704)	3.403
White							-2.162(.863)	.115*
ShareFriends							.395(.248)	1.484
RespectFriends							-.272(.146)	.762
Nag							.366(.109)	1.442**
Constant	-4.937(.653)	.007***	-4.890(.575)	.008***	-5.415(.691)	.004***	-7.372(2.367)	.001**
-2 Log-likelihood	111.054		100.680		96.803		74.926	
Model Chi-Square	8.530*		18.904***		22.782***		44.658***	
Cox & Snell R^2	.018		.038		.046		.088	
Nagelkerke R^2	.080		.175		.210		.303	

* $p < .05$, ** $p < .01$, *** $p < .001$

Variables retained at the .20 level were found to explain 4.6% to 21.0% of the variation in the dependent variable. Providing personal information to online contacts (ProvidedInfo) continued to increase the likelihood of participating in a sexual encounter with an online contact (b = .283, p < .001). In addition to this variable, having a person designated as "Other" in the room during Internet use (OthInRm) also increased the likelihood of a sexual encounter by almost 4 times [Exp(B) = 3.707].

The full logistic regression model, including the control variables, is presented in Table 40. Variables retained at the .20 level were shown to explain 8.8% to 30.3% of the variation in the dependent variable. Greater hours per week use of instant messaging (b = .152, p < .05) was found to increase the likelihood of participating in a sexual encounter with a person met online. Moreover, providing personal information to online contacts (ProvidedInfo) (b = .410, p < .001) remained as a statistically significant positive predictor of participating in a sexual.encounter with a person met online. Furthermore, respondents who reported parents nagged them more often (Nag) were significantly more likely to participate in offline contact with a person met online (b = .366, p < .01). Last, respondents who were white (White) were less likely than nonwhites to participate in a sexual encounter with an online contact (b = -2.162, p < .05)

Split Models

Table 41 presents the logistic regression estimates for males and females for the dependent variable "receipt of unwanted sexually explicit material" during the college freshman time period. Only the full models are presented; stepwise regression with backward elimination was used in the analysis, with the entire set of independent and control variables initially inserted. Variables retained at the .20 level were shown to explain 12.9% to 25.8% of the variation in the dependent variable for males, and 11.3% to 17.6% for females. Similar to the high school senior split model, males and females did not share any statistically significant predictors. The males in the sample did have three statistically significant predictors of the receipt of sexually explicit material. Participating in website design (Design) (b = 1.445, p < .05) and having a teacher in the room during Internet use

Table 41. Logistic Regression Estimates for the Dependent Variable Receipt of Sexually Explicit Material During the College Freshman Time Period

Variable	Male Model (N = 195)		Variable	Female Model (N = 288)	
	B(SE)	Exp(B)		B(SE)	Exp(B)
Design	1.445(.637)	4.242*	Research	-3.467(1.288)	.031**
LivRm	-1.077(.580)	.341	Facebook	-2.118(.905)	.120*
TeachInRm	1.279(.643)	3.594*	SNWInfo	.393(.160)	1.482*
SiblnRm	1.017(.553)	2.766	Comm	1.377(.537)	3.962*
Dorm	-.925(.653)	.397	OthInRm	1.874(.532)	6.515***
ShareParents	.285(.184)	1.329	ActMon	1.894(1.079)	6.646
EnjoyFriends	.393(.288)	1.481	White	1.057(.790)	2.877
RespectParents	-.450(.196)	.638*	ShareParents	-.163(.107)	.850
Participate	.161(.104)	1.175	RespectParents	.285(.185)	1.331
Privileges	-.331(.194)	.718	Succeed	.517(.286)	1.677
Constant	-4.514(2.473)	.011	Yell	.229(.095)	1.257*
			Constant	-7.796(2.759)	.000**
-2 Log-likelihood	106.232		-2 Log-likelihood	262.785	
Model Chi-Square	78.895***		Model Chi-Square	34.626***	
Cox & Snell R^2	.129		Cox & Snell R^2	.113	
Nagelkerke R^2	.258		Nagelkerke R^2	.176	

* $p < .05$, ** $p < .01$, *** $p < .001$

(TeachInRm) (b = 1.279, p < .05) increased the likelihood that the males in the sample would receive unwanted sexually explicit material online. Conversely, males who respected their parents (RespectParents) were less likely to be victimized (b = -.450, p < .05).

Females, alternatively, had a number of statistically significant predictors of the dependent variable. Female respondents who had a person designated as "Other" in the room with them during Internet use (OthInRm) were more than 6 times more likely to receive unwanted sexual material [Exp(B) = 6.515]. Providing personal information on asocial networking website (SNWInfo) (b = .393, p < .05), communicating with others online (Comm) (b = 1.377, p < .05), and being yelled at by parents (Yell) (b = .229, p < .05) were also factors that increased the likelihood of receiving unwanted sexually explicit material online. However, behaviors such as researching online (Research) (b = -3.467, p < .01) and using Facebook (b = -2.118, p < .05) decreased the likelihood of receipt of unwanted sexually explicit material.

The logistic regression estimates for males and females for the dependent variable "receipt of non-sexual harassment" during the college freshman time period are presented in Table 42. The variables retained at the .20 level again were shown to explain only 7.5% to 16.9% of the variation in the dependent variable for males, and 9.7% to 16.3% for females. Males had only one statistically significant predictor of the dependent variable; respondents who provided personal information to online contacts (ProvidedInfo) had a greater likelihood of victimization (b = .171, p < .05). With regard to females, communicating with others online (Comm) (b = .920, p < .01) and enjoying spending time with friends (EnjoyFriends) (b = .353, p < .01) were two variables that increased the likelihood of receipt of non-sexual harassment. Alternatively, sharing thoughts and feelings with friends (ShareFriends) decreased the likelihood of victimization for females (b = -.320, p < .01).

Table 43 presents the logistic regression estimates for males and females for the dependent variable "receipt of sexual solicitation" during the college freshman time period. Variables retained at the .20 level were shown to explain 15.0% to 38.7% of the variation in the dependent variable for males, but only 5.8% to 17.3% for females. Males and females again shared no statistically significant predictors. With regard to males, respondents who had a person designated as "Other" in the room with them during Internet use (OthInRm) were more likely to receive sexual

Table 42. Logistic Regression Estimates for the Dependent Variable Receipt of Non-Sexual Harassment During the College Freshman Time Period (N=483)

Variable	Male Model (N = 195)		Variable	Female Model (N = 288)	
	B(SE)	Exp(B)		B(SE)	Exp(B)
IntHours	.035(.020)	1.036	Travel	.640(.347)	1.897
Shop	.781(.615)	2.184	Design	.699(.432)	2.011
ProvidedInfo	.171(.087)	1.187*	Comm	.920(.348)	2.508**
FiltSoft	.836(.556)	2.306	RestrictAdult	-1.282(.798)	.277
Age	-.824(.567)	.439	ShareParents	.189(.088)	1.208*
Grades	.396(.257)	1.487	ShareFriends	-.320(.122)	.726**
Constant	7.422(10.316)	1672.028	EnjoyFriends	.353(.180)	1.423*
			Nag	.138(.065)	1.147*
			Constant	-4.609(1.512)	.010**
-2 Log-likelihood	100.123		-2 Log-likelihood	229.812	
Model Chi-Square	15.302*		Model Chi-Square	29.345***	
Cox & Snell R^2	.075		Cox & Snell R^2	.097	
Nagelkerke R^2	.169		Nagelkerke R^2	.163	

* p < .05; ** p < .01; *** p < .001

187

Table 43. Logistic Regression Estimates for the Dependent Variable Receipt of Sexual Solicitation During the College Freshman Time Period (N = 483)

Variable	Male Model (N = 195)		Variable	Female Model (N = 288)	
B(SE)	Exp(B)		B(SE)	Exp(B)	
OthInRm	2.371(.806)	10.706**	Design	1.232(.622)	3.427*
Age	1.565(.891)	4.783	Comm	1.069(.599)	2.912
LivDorm	2.124(1.390)	8.365	RestrictOther	2.334(1.007)	10.324*
GPA	-.656(.346)	.519	DKActMon	1.611(.685)	5.008*
ShareParents	.483(.292)	1.621	DKFiltSoft	-2.619(1.375)	.073
ShareFriends	-.744(.258)	.475**	GPA	.429(.288)	1.536
EnjoyFriends	1.125(.483)	3.081*	Constant	-4.758(.827)	.009***
RespectParents	-.467(.246)	.627			
Grades	.983(.457)	2.673*			
Succeed	-.780(.306)	.458*			
Yell	.267(.144)	1.307			
Constant	-40.485(18.410)	.000*			
-2 Log-likelihood	63.812		-2 Log-likelihood	100.568	
Model Chi-Square	31.572**		Model Chi-Square	17.284**	
Cox & Snell R²	.150		Cox & Snell R²	.058	
Nagelkerke R²	.387		Nagelkerke R²	.173	

* p < .05; ** p < .01; *** p < .001

solicitation (b = 2.371, p < .01), as well as those who enjoyed spending time with friends (EnjoyFriends) (b = 1.125, p < .05). Conversely, sharing thoughts and feelings with friends (ShareFriends) (b = -.475, p < .01) decreased the likelihood of victimization for males. Placing an importance on grades (Grades) (b = .983, p < .05), as well as the desire to succeed academically (Succeed) (b = -.780, p < .05), were also significant predictors of sexual solicitation.

Females had only a few significant predictors of sexual solicitation. First, females who participated in website design while online (Design) (b = 1.232, p < .05), as well as if being unsure whether they were being actively monitored while using the Internet (DKActMon) (b = 1.611, p < .05), were more likely to be victimized. Also, having restrictions on Internet activities designated as "Other" (RestrictOther) (b = 2.334, p < .05) increased the likelihood of sexual solicitation.

The remaining dependent variables examined the formation of various types of relationships with online contacts. Table 44 presents the logistic regression estimates for males and females for the dependent variable "formation of a virtual relationship with an online contact" during the college freshman time period. The variables retained at the .20 level were shown to explain a very respectable 40.3% to 64.7% of the variation in the dependent variable for males, but a lower 21.8% to 35.0% for females.

Males and females shared two statistically significant predictors, which were similar to the total sample model. Both male (b = .503, p < .001) and female (b = .253, p < .05) respondents who provided personal information to online contacts (ProvidedInfo) had a greater likelihood of formation of a virtual relationship with an online contact, as did males (b = 3.288, p < .001) and females (b = 1.026, p < .05) who communicated with online contacts (Comm). Participation in these behaviors appeared to have a greater impact on males, as providing personal information to online contacts increased the occurrence of forming a virtual relationship by 65% for each type of information provided [Exp(B) = 1.654], and communicating with online contacts increased the likelihood of the dependent variable by almost 17 times [Exp(B) = 16.792].

Table 44. Logistic Regression Estimates for the Dependent Variable Formation of Virtual Relationship During the College Freshman Time Period (N = 483)

Variable	Male Model (N = 195)		Variable	Female Model (N = 288)	
	B(SE)	Exp(B)		B(SE)	Exp(B)
Shop	-1.216(.596)	.296*	IntHours	.024(.013)	1.025
IMHours	-.170(.074)	.843*	Design	.684(.465)	1.981
SNWHours	.314(.082)	1.369***	Social	1.415(1.083)	4.117
NoPriv	2.684(.759)	14.651***	ChatHour	2.340(.860)	10.365**
Comm	3.288(.917)	16.792***	NoPriv	.492(.376)	1.635
ProvidedInfo	.503(.125)	1.654***	Comm	1.026(.396)	2.789*
Dorm	1.648(.944)	5.195	ProvidedInfo	.253(.081)	1.288*
NoOneRm	.988(.593)	2.686	Friend	-.570(.411)	.566
RestrictOther	-4.722(2.323)	.009*	RestrictAdult	-1.138(.822)	.320
DKActMon	1.619(.706)	5.046*	Constant	-4.152(1.191)	.016***
EnjoyFriends	-.336(.225)	.714			
RespectParents	.621(.281)	1.860*			
RespectTeach	-.870(.276)	.419**			
Yell	-.416(.242)	.660			
Privileges	.499(.257)	1.647			
Constant	-4.298(2.550)	.014			
-2 Log-likelihood	88.980		-2 Log-likelihood	210.106	
Model Chi-Square	100.070***		Model Chi-Square	70.768****	
Cox & Snell R^2	.403		Cox & Snell R^2	.218	
Nagelkerke R^2	.647		Nagelkerke R^2	.350	

* $p < .05$, ** $p < .01$, *** $p < .001$

Males and females also had unique statistically significant predictors of formation of a virtual relationship. Specifically with regard to the males in the sample, increased use of social networking websites (SNWHours) (b = .314, p < .001) and use of a non-privatized social networking website (NoPriv) (b = 2.684, p < .001) significantly increased the likelihood of forming a virtual relationship. Other significant predictors of the dependent variable for males were as follows: shopping online (b = -1.216, p < .05); use of instant messaging (b = -.170, p < .05); restrictions on Internet use designated as "Other" (b = -4.722, p , .05); being unsure if actively monitored by parents/ guardians during Internet use (DKActMon) (b = 1.619; p < .05); having respect for parents (RespectParents) (b = .621, p < .05); and having respect for teachers (RespectTeachers) (b = -.870, p < .01). Alternatively, females who used chat rooms (ChatHour) also experienced an increased likelihood of forming a virtual relationship with an online contact (b = 2.340, p < .01).

The logistic regression estimates for males and females of "participation in offline contact" during the college freshman time period are presented in Table 45. Variables retained at the .20 level were shown to explain 31.3% to 49.8% of the variation in the dependent variable for males, and 21.8% to 39.1% for females.
Males and females had one shared statistically significant predictor. Males (b = 3.429, p < .001) and females (b = 1.039, p < .05) who communicated with online contacts (Comm) were more likely to participate in offline contact with a person met online.

Males and females also had several unique statistically significant predictors. Males who were nagged by parents (Nag) (b = .298, p < .01) and reported greater hours per week use of social networking websites (SNWHours) (b = .210, p < .001) experienced an increased likelihood of participation in offline contact. Moreover, having a person designated as "Other" in the room during Internet use (OthInRm) (b = 1.794, p < .01), as well as having privileges revoked by parents (Privileges) (b = .347, p < .05), also increased the likelihood of participating in an offline relationship for males. Greater yelling by parents (Yell) (b = -.417, p < .05) was shown to decrease the likelihood of participation in offline contact. With regard to female respondents, providing one or more types of personal information to online contacts (ProvidedInfo) increased the likelihood of participating in an offline relationship (b =

Table 45. Logistic Regression Estimates for the Dependent Variable Participation in Offline Contact During the College Freshman Time Period (N = 483)

Variable	Male Model (N = 195)		Variable	Female Model (N = 288)	
	B(SE)	Exp(B)		B(SE)	Exp(B)
Shop	-.657(.482)	.518	MySpace	.743(.469)	2.102
SNWHours	.210(.056)	1.233***	NoPriv	-1.429(.553)	.239*
Comm	3.429(.738)	10.580***	Comm	1.039(.483)	2.826*
OthInRm	1.794(.606)	6.013**	ProvidedInfo	.477(.098)	1.612***
LivDorm	-1.105(.631)	.331	LivOther	-2.191(1.805)	.112
ShareFriends	.335(.192)	1.398	GPA	-.320(.245)	.726
EnjoyFriends	-.288(.216)	.750	Privileges	.217(.120)	1.242
RespectTeach	-.261(.195)	.770	Constant	-2.920(.583)	.054***
Yell	-.417(.190)	.659*			
Nag	.298(.108)	1.347**			
Privileges	.347(.171)	1.414*			
Constant	-3.037(1.753)	.048			
-2 Log-likelihood	119.101		-2 Log-likelihood	164.726	
Model Chi-Square	72.816***		Model Chi-Square	70.989***	
Cox & Snell R^2	.313		Cox & Snell R^2	.218	
Nagelkerke R^2	.498		Nagelkerke R^2	.391	

* $p < .05$; ** $p < .01$; *** $p < .001$

.477, p < .001). Alternatively, use of a non-privatized social networking website (NoPriv) by females decreased the likelihood of participating in offline contact with a person met online (b = -1.429, p < .05).

Table 46 presents the logistic regression estimates for males and females for the dependent variable "face-to-face contact with an online contact" during the college freshman time period. Variables retained at the .20 level were shown to explain a healthy 26.2% to 46.1% of the variation in the dependent variable for males, and a similar 21.9% to 48.2% for females. Both male (b = .232, p < .01) and female (b = .413, p < .001) respondents who provided personal information to online contacts (ProvidedInfo) had a greater likelihood of participating in face-to-face contact with an online contact, as well as males (b = 2.742, p < .01) and females (b = 1.597, p < .05) who communicated with online contacts (Comm).

There were also separate statistically significant predictors for males and females. Males who had greater hours per week usage of social networking websites (SNWHours) (b = .125, p < .05), as well as those having a person designated as "Other" in the room during Internet use (OthInRm) (b = 1.269, p < .05) and those often nagged by parents (Nag) (b = .209, p < .05) were more likely to meet an online contact face-to-face.

With regard to females, those who had greater hours per week use of the Internet (IntHours) (b = .085, p < .001), had a friend in the room during Internet use (FriInRm(b = 1.801), and respected teachers (RespectTeachers) (b = .585, p < .05), were more likely to participate in face-to-face contact with an online contact. Conversely, females who more often shared thoughts and feelings with friends (ShareFriends) (b = -.386, p < .01), and those who respected their parents (RespectParents) (b = -.501, p < .01) and had a desire to succeed in school (Succeed) (b = -.869, p < .01) were less likely to participate in face-to-face contact with a person met online.

The last logistic regression estimates for males and females were for the dependent variable "sexual encounter with an online contact" during the college freshman time period, as presented in Table 47. Variables retained at the .20 level were shown to explain 15.4% to 37.7% of the variation in the dependent variable for males, but only 6.8% to 27.1% for females. Males and females had one shared statistically significant predictor. Males who provided personal

Table 46. Logistic Regression Estimates for the Dependent Variable Face to Face Contact During the College Freshman Time Period (N = 483)

Variable	Male Model (N = 195) B(SE)	Exp(B)	Variable	Female Model (N = 288) B(SE)	Exp(B)
Shop	-1.008(.537)	.365	IntHours	.085(.022)	1.088***
SNWHours	.125(.055)	1.133*	NoPriv	-1.469(.681)	.230*
Comm	2.742(.834)	15.524**	Comm	1.597(.680)	4.939*
ProvidedInfo	.232(.084)	1.261**	ProvidedInfo	.413(.109)	1.512***
OthInRm	1.269(.595)	3.557*	FriInRm	1.801(.788)	6.056*
LivOther	1.643(.875)	5.169	GPA	-.439(.299)	.644
EnjoyFriends	-.300(.176)	.741	ShareFriends	-.386(.137)	.680**
Participate	.152(.106)	1.164	RespectParents	-.501(.162)	.606**
Nag	.209(.089)	1.232*	RespectTeach	.585(.272)	1.795*
Constant	-4.197(1.640)	.015*	Grades	.454(.322)	1.575
			Succeed	-.869(.310)	.419**
			Constant	-1.323(.461)	.266
-2 Log-likelihood	104.641		-2 Log-likelihood	103.353	
Model Chi-Square	59.346***		Model Chi-Square	71.279***	
Cox & Snell R^2	.262		Cox & Snell R^2	.219	
Nagelkerke R^2	.461		Nagelkerke R^2	.482	

* $p < .05$; ** $p < .01$; *** $p < .001$

Table 47. Logistic Regression Estimates for the Dependent Variable Sexual Encounter During College Freshman Time Period (N = 483)

Variable	Male Model (N = 195) B(SE)	Exp(B)	Variable	Female Model (N = 288) B(SE)	Exp(B)
IntHours	-.221(.096)	.802*	IntHours	.066(.035)	1.068
EmHours	.742(.438)	2.101	ProvidedInfo	.314(.139)	1.369*
IMHours	.407(.189)	1.502	White	-1.783(1.065)	.168
ProvidedInfo	.631(.217)	1.879**	RespectTeach	.641(.465)	1.898
FriInRm	-1.612(1.269)	.200	Participate	-.243(.171)	.784
OthInRm	3.617(1.528)	17.215*	Nag	.337(.159)	1.401*
Constant	-5.747(1.848)	.003***	Constant	-10.797(4.218)	.000**
-2 Log-likelihood	27.814		-2 Log-likelihood	38.068	
Model Chi-Square	32.437***		Model Chi-Square	20.260**	
Cox & Snell R²	.154		Cox & Snell R²	.068	
Nagelkerke R²	.377		Nagelkerke R²	.271	

*p < .05; ** p < .01; *** p < .001

information to online contacts (ProvidedInfo) were significantly more likely to participate in a sexual encounter with a person met online (b = .631, p < .01), and females were also more likely to do the same (b = .314, p < .05). With regard to males, greater hours per week use of the Internet (IntHours) decreased the likelihood of a sexual encounter (b = -.221, p < .05), but having a person designated as "Other" in the room during Internet use (OthInRm) increased this likelihood (b = 3.617, p < .05). The only other significant predictor of participation in a sexual encounter for females was being nagged by parents. Females who were nagged more often by parents were more likely to participate in a sexual encounter with a person met online (b = .337, p < .05).

HYPOTHESES

The current research sought to add to the existing body of literature by testing seven hypotheses focusing on adolescent online victimization and relationship formation. The results of the multivariate analysis presented in this chapter provide evidence to consider in assessing whether the original hypotheses are supported.

Hypothesis 1

The first hypothesis of the study was based on the theoretical construct of exposure to motivated offenders, as derived from Routine Activities Theory, and it stated the following: Adolescents who spend more time on the Internet using modes of computer-mediated communication are more likely to be victimized online and form relationships with online contacts. After thorough analysis of the data from the high school senior and college freshman time periods, it can be concluded that at least some support was found for this hypothesis.

With regard to the high school senior time period, there were several independent variables measuring exposure to motivated offenders that were shown to increase the likelihood of two of the three types of victimization: receipt of unwanted sexually explicit material and non-sexual harassment. Shopping online and using chat rooms increased the likelihood of receipt of unwanted sexually explicit material, while using the Internet to socialize and greater hours per week use of email was shown to increase the likelihood of receipt of non-sexual harassment. However, when comparing the sexes, only males had variables measuring exposure to motivated offenders emerge as significant predictors of victimization. Males who used chat rooms were more likely to receive sexually explicit material, and males who used the Internet to socialize were more likely to receive non-sexual harassment.

Although not direct forms of victimization, other dependent variables measuring relationship formation were influenced by variables measuring exposure to motivated offenders during the high school senior time period. For example, respondents who used MySpace were more likely to participate in face-to-face contact with a person met online. Males who used Facebook, and females who used MySpace, were more likely to participate in a virtual relationship with a person met online. Moreover, males who spent increased hours per week using social networking websites were more likely to participate in offline contact and more likely to experience a sexual encounter with a person met online.

Concerning the college freshman time period, there were several independent variables measuring exposure to motivated offenders that were shown to increase the likelihood of the three types of victimization. Using chat rooms increased the likelihood of receipt of unwanted sexually explicit material, while using the Internet to plan travel was shown to increase the likelihood of receipt of non-sexual harassment. When comparing the sexes, participating in website design increased the likelihood that males would receive sexually explicit material and females would receive sexual solicitation.

Other dependent variables measuring relationships were influenced by variables measuring exposure to motivated offenders during the college freshman time period. For example, respondents who spent greater hours per week on the Internet were more likely to participate in a virtual relationship, offline contact, and face-to-face contact with a person met online. Males who spent greater hours per week using social networking websites were more likely to participate in a virtual relationship, as well as offline contact and face-to-face contact with a person met online. Females who used chat rooms were more likely to participate in a virtual relationship, and those who spent greater hours per week on the Internet were more likely to participate in a sexual encounter with a person met online.

Hypothesis 2

The second hypothesis of the study was based on the theoretical construct of target suitability and stated the following: Adolescents who provide personal information to online contacts are more likely to be victimized online. Based on the results from the high school senior and college freshman time periods, it can be concluded that fairly strong support was found for this hypothesis.

With regard to the high school senior time period, providing personal information was consistently found to increase the likelihood of the three types of victimization, for both the entire sample and by specific gender. Providing more types of various pieces of personal information to online contacts was shown to increase the likelihood of receipt of non-sexual harassment and sexual solicitation for the entire sample, as well as for males and females separately. Each gender also had unique predictors of victimization. Female respondents who provided personal information to online contacts were more likely to receive unwanted sexually explicit material, and those who posted personal information on a social networking website were more likely to receive sexual solicitation. For males, however, there was only one target suitability variable that was shown to increase the likelihood of one type of victimization during the college freshman time period: providing personal information to online contacts was found to increase the likelihood of receipt of non-sexual harassment.

Hypothesis 3

The third hypothesis of the study also was based on the theoretical construct of target suitability and stated the following: Adolescents who do not privatize their social networking websites for viewing by only approved online contacts are more likely to be victimized online. After thorough analysis of the data from the high school senior and college freshman time periods, it can be concluded that there was little support found for this hypothesis.

For both the high school senior and college freshman time periods, use of a non-privatized social networking website was not found to be a statistically significant predictor of victimization. The only dependent variable that was affected by this activity was formation of a virtual relationship. According to the results, respondents who used a non-privatized social networking website were more likely to form a virtual relationship with a person they met online.

Hypothesis 4

The fourth hypothesis of the study again was based on the theoretical construct of target suitability and stated the following: Adolescents who provide personal information to online contacts are more likely to form offline relationships with contacted individuals. Based on the results from the high school senior and college freshman time periods, it can be concluded that solid support was found for the hypothesis.

Overall, for the high school senior time period, there were several target suitability variables that were found to increase the likelihood of forming relationships with people met online. Communicating with people online and providing personal information to online contacts were both shown to consistently increase the likelihood of forming virtual relationships, participating in offline contact, and meeting online contacts face-to-face. This was true for the entire sample and for males and females separately. Providing personal information to online contacts also increased the likelihood of participating in a sexual encounter with an online contact for the entire sample, and particularly for females when they were examined separately.

As for the college freshman time period, there again were several independent variables measuring target suitability that were revealed to increase the likelihood of forming relationships with people met online. Communicating with people online and providing personal information to online contacts were both found to increase the likelihood of forming virtual relationships, participating in offline contact, and meeting online contacts face-to-face for the entire sample. For both males and females, these two variables had positive influences on forming virtual relationships and participating in offline contact. Providing information to online contacts also increased the likelihood of participating in a sexual encounter with an online contact for the entire sample, as well as for both males and females when examined separately.

Hypothesis 5

The fifth hypothesis of the study was based on the theoretical construct of lack of capable guardianship and stated the following: Adolescents who utilize protective software are less likely to be victimized online. Based on data pertaining to the high school senior and college freshman time periods, it can be concluded that little support was found for this hypothesis.

The only statistically significant predictors of victimization with regard to protective software were observed when respondents were unsure if it is was present, and in those cases it was a positive predictor of victimization. With regard to the high school senior time period, it was shown that male respondents who were unsure if filtering software was installed on the computer were more likely to receive unwanted sexually explicit material. Although not a direct form of victimization, being unsure of the presence of filtering and blocking software also was shown to be a significant predictor of formation of a virtual relationship. During the high school senior time period, for the entire sample as well as females when examined separately, respondents were more likely to form a virtual relationship with an online contact if they were unsure whether protective software was present on their computers.

Hypothesis 6

The sixth hypothesis of the study also was based on the theoretical construct of lack of capable guardianship and stated the following: Adolescents who have restricted use of the Internet are less likely to be victimized online. Based on the results from the high school senior and college freshman time periods, it can be concluded that limited support was found for this hypothesis.

With regard to the high school senior time period, the results indicated that males who had restrictions on viewing adult websites were less likely to receive unwanted sexually explicit material and sexual solicitation. However, there were no statistically significant predictors for the college freshman time period. Other dependent variables measuring relationships were influenced by variables measuring restricted use of the Internet. Respondents who had restrictions on viewing adult websites were less likely to participate in face-to-face contact with a person met online, and females who had restrictions on viewing adult websites were less likely to participate in offline contact with a person met online. On the other hand, females who had restrictions on use of CMCs were more likely to participate in offline contact and face-to-face contact with a person met online. This again may be an issue of temporal ordering, as respondents who had participated in offline contact may then have had restrictions placed on their use of CMCs.

Hypothesis 7

The final hypothesis of the study was based on the theoretical construct of lack of capable guardianship and stated the following: Adolescents who are monitored while using the Internet are less likely to be victimized online. After thorough analysis of the data from the high school senior and college freshman time periods, it can be concluded that mixed support was found for this hypothesis.

Concerning the high school senior time period, females who reported having a teacher in the room during Internet use were less likely to receive non-sexual harassment. In contrast, males who had a parent in the room during Internet use were more likely to receive unwanted sexually explicit material online. As mentioned in the previous section, though, an issue with temporal ordering may be present in this latter relationship as well.

With regard to the dependent variables representing relationship formation, females who had a friend in the room during Internet use were less likely to form a virtual relationship, participate in offline contact or face-to-face contact, or participate in sexual encounter with a person met online. Males who had a parent in the room also were less likely to participate in offline contact with a person met online.

Finally, with regard to the college freshman time period, females who had a teacher in the room during Internet use again were less likely to receive non-sexual harassment. Respondents who had a friend in the room also were less likely to receive unwanted sexually explicit material. However, respondents with a person designated as "Other" in the room with them during Internet use were more likely to receive unwanted sexually explicit material and sexual solicitation, and were more likely to experience offline and face-to-face contacts with a person met online. Furthermore, males with a parent in the room were more likely to participate in face-to-face contact with a person met online, taking into account a possible issue with temporal ordering.

SUMMARY

The purpose of this chapter was to provide a presentation of the multivariate results from the analysis of data collected from 483 college freshmen at a mid-sized university in the northeast. Data originated from surveys administered to the respondents during the 2008 spring semester in freshman-level classes. Hypotheses were previously derived from the theoretical constructs of Routine Activities Theory, and data were collected on the types of Internet behaviors respondents participated in while a high school senior and college freshman.

Results from the high school senior time period were presented first. Upon examination of the total sample, shopping online, use of chat rooms, and providing personal information to online contacts were shown to be significant positive predictors of receipt of unwanted sexually explicit material online. Socializing online, greater email hours, and providing personal information to online contacts were shown to increase the likelihood of receipt of non-sexual harassment. Providing personal information to online contacts was also a significant predictor of receipt of sexual solicitation, along with using the Internet in a location designated as "Other." The results also indicated that respondents whose parents took away privileges were more likely to receive unwanted sexually explicit material and sexual solicitation. As noted previously in the text, there could be an issue of temporal ordering present.

With regard to the dependent variables representing the formation of relationships with online contacts, there were two variables shown to be consistent significant predictors. Respondents who communicated with people online and provided personal information to online contacts were more likely to form a virtual relationship, participate in offline contact, and participate in face-to-face contact with a person met online. Much like the dependent variable measuring sexual solicitation, providing information to online contacts was also a significant positive predictor of participating in a sexual encounter with a person met online.

Split models also were examined for the high school senior time period to consider possible differences in predictors between males and females. With regard to the three types of victimization, males and females shared some significant predictors for two dependent variables. Both males and females who provided personal information to online contacts were shown to be more likely to receive non-sexual harassment and sexual solicitation. However, there were no similarities in predictors of receipt of unwanted sexually explicit material. To illustrate, males who used chat rooms, Facebook, and school computer labs were more likely to receive unwanted sexual material. In contrast, females who provided personal information to online contacts were more likely to receive this unwanted material.

As for the dependent variables representing formation of relationships with online contacts, similar to the models examining the entire sample, providing personal information to online contacts and communicating with people online were significant predictors of formation of a virtual relationship, participating in offline contact with a person met online, and meeting an online contact face-to-face for both males and females. Females who provided personal information to online contacts were also more likely to participate in a sexual encounter with a person met online. In addition, males who used Facebook were shown to be more likely to form a virtual relationship and participate in offline contact with a person met online.

Results from the college freshman time period then were examined. Based on examination of the entire sample, having a person in the room designated as "Other" and use of chat rooms were shown to be the most significant positive predictors of receipt of unwanted sexually explicit material online, while planning travel online and communicating with others online were shown to increase the likelihood of receipt of non-sexual harassment. As college freshmen, females were 3 times more likely than males to receive non-sexual harassment. In addition, communicating with others online was also a significant predictor of receipt of sexual solicitation.

Concerning the dependent variables representing the formation of relationships with online contacts while a college freshman, there were two variables again shown to be consistent significant predictors. Respondents who communicated with others online and those who provided personal information to online contacts were more likely to form a virtual relationship, participate in offline contact, and participate in face-to-face contact with a person met online. Furthermore, providing personal information to online contacts was also a significant predictor of participating in a sexual encounter, much during the high school senior time period.

Split models also were examined for the college freshman time period to examine possible differences between males and females. Both males and females who provided personal information to online contacts and communicated with others online were more likely to receive non-sexual harassment, and females who participated in those same behaviors were more likely to receive unwanted sexually explicit material online. There were no similarities in predictors of receipt of sexual solicitation. For instance, males who had a person designated as "Other" in the room during Internet use and females who participated in website design were more likely to receive sexual solicitation.

As for the dependent variables representing formation of relationships with online contacts, similar to the models examining the entire sample, providing personal information to online contacts and communicating with people online were significant predictors of formation of a virtual relationship, participating in offline contact, and meeting an online contact face-to-face for both males and females. Males who provided personal information to online contacts were also more likely to participate in a sexual encounter with a person met online. In addition, males who reported greater hours per week of social networking website use were more likely to form a virtual relationship and participate in face-to-face contact with a person met online.

Finally, after thorough review of the results of the models encompassing the two time periods examined in this study, the original hypotheses were assessed. Based on the findings, solid support was found for Hypotheses 1, 2, and 4. Based on the hypothesis tests, it can be concluded that adolescents who spend more time on the Internet using modes of computer-mediated communication and those who provide personal information to online contacts are more likely to be victimized online. Also, adolescents who provide personal information to online contacts are more likely to form virtual and offline relationships with those contacts. Mixed support was found for Hypothesis 6, which stated that adolescents who have restricted use of the Internet are less likely to be victimized online, and Hypothesis 7, which stated that adolescents who are monitored while using the Internet are less likely to be victimized online. Little support was found for Hypothesis 3, which stated that adolescents who do not privatize their social networking websites for viewing by only approved online contacts are more likely to be victimized online, as well as Hypothesis 5, which stated that adolescents who use protective software are less likely to be victimized online.

The next and final chapter will present a discussion of the conclusions drawn from this study. First, a comparison of the findings from this study versus those from relevant past research will be presented. Next, policy implications based on this body of research will be proposed. Finally, the limitations of this study and recommendations for future research will be discussed.

CHAPTER 7:

Applying the Findings to the Future

Daily use of the Internet is now common for many Americans, whether for socialization, research, or various other activities. Considering the fact that the idea of the Internet was not conceived until 1962 (Leiner et al., 2003), and computers only started to be used in many businesses and homes in the early 1990s (Sanger et al., 2004), this commodity of communication that has become a prevalent mainstay in American homes has been a rapidly growing contemporary development. Due to its current availability and easy accessibility, the frequency of Internet use has increased for all age groups. However, Internet use by adolescents has experienced the largest increase, as compared to any other age group (Addison, 2001).

Today's adolescents grew up using the Internet; in turn, they are very familiar with the multiple opportunities available online. Youth are especially involved in online socialization, using various methods of computer-mediated communication (CMC), such as email, chat rooms, instant messaging, and social networking websites. Not only are more adolescents using the Internet to socialize, but they are also spending more time online (Izenberg & Lieberman, 1998; Nie & Erbring, 2000; United States Department of Commerce, 2002). Unfortunately, while the use of CMCs can produce positive interaction and the development of enjoyable relationships, young people spending extensive amounts of time online also may be at greater risk for victimization.

Several past studies of Internet use by adolescents have found that increasing numbers of young people are experiencing the following types of victimization while online: unwanted exposure to sexually explicit material, sexual solicitation, and non-sexual harassment (Mitchell et al., 2003; Mitchell, Finkelhor, & Wolak, 2007; O'Connell et al., 2002; Quayle & Taylor, 2003; Sanger et al., 2004; Wolak et al., 2002; Wolak et al., 2003; Wolak et al., 2004; Wolak et al., 2006; Wolak, Mitchell, & Finkelhor, 2007; Ybarra, Mitchell, Finkelhor, & Wolak, 2007). The majority of these studies have been descriptive in nature, and therefore, there has been a lack of explanatory research that indicates what online behaviors and activities may increase (or decrease) the likelihood of victimization. Of the few explanatory studies performed, use of chat rooms, discussion of sexual topics with online contacts, and a tumultuous relationship with family or friends have been noted to increase the odds of online victimization (Mitchell et al., 2007; Wolak et al., 2007; Ybarra et al., 2007).

The purpose of this study was to further investigate Internet use among a sample of college freshmen and to consider their experiences with online victimization and the formation of relationships with online contacts. In order to more fully examine this topic area, the chosen methodology was developed under the concepts and propositions of Routine Activities Theory, which has been utilized many times in the past to explain various types of victimization. This study employed a survey and was anticipated to produce a more complete understanding of adolescent Internet use and victimization.

The remainder of this chapter will discuss the major findings and conclusions of the study. First, the effects of Internet behaviors and activities on online victimization and the formation of relationships with online contacts will be discussed. Next, a discussion of relevant policy implications will be presented, followed by the limitations of the study and suggestions for future research. Finally, some concluding comments will be provided.

ROUTINE ACTIVITIES THEORY AND ONLINE VICTIMIZATION OF ADOLESCENTS

This study utilized data that were collected from 483 college freshmen at a mid-sized university in the northeast, based on surveys administered to the respondents during the 2008 spring semester in freshman-level classes. In general, the online behaviors of this sample were similar to those of samples of respondents who participated in previous studies of online victimization. For instance, with regard to the use of CMCs, a large majority of respondents in this study reported frequent use of email, instant messaging, and social networking websites. However, contrary to past studies, less than 10% of this sample reported the use of chat rooms during both the high school senior and college freshman time periods. Previous studies of adolescent Internet use revealed a much higher percentage of respondents who reported the use of chat rooms (Lenhart et al., 2001; Sanger et al., 2004; Wolak et al., 2002). Analysis of the data obtained from the college freshmen was separated into three sections. Univariate results focused on the frequencies of the variables and descriptive statistics for the sample. Bivariate results revealed the statistical associations among the independent variables and dependent variables. Finally, multivariate results were used to further assess relationships between the independent variables, representing the theoretical constructs of Routine Activities Theories, and the dependent variables, representing online victimization and the formation of relationships with online contacts, while controlling for other factors. Although there is a rather limited amount of past research on the topic of adolescent online use and victimization, the findings of this study can be compared to the results of previous research, as well as other studies of victimization that have tested Routine Activities Theory.

UNIVARIATE RESULTS

Univariate results were separated by time period in question (i.e., high school senior and college freshman), as well as each theoretical construct. This approach provided a presentation of the results of the frequencies and descriptive statistics for each variable measured in the survey. The first theoretical construct examined was exposure motivated offenders, which entailed measuring the types of activities performed online and the amount of time spent using the Internet. Overall, during the college freshmen time period, respondents reported spending more time on the Internet (4.21 hours per day and 6.49 days per week) compared to the high school time period (2.85 hours per day and 5.92 days per week). Although a larger percentage of respondents reported using email, instant messaging, and social networking websites during the college freshman time period, a comparable 80% or more of the respondents also reported that during the high school senior time period they utilized these CMCs.

The next section of the survey measured respondents' Internet use that potentially increased target suitability. During the high school time period, a substantial percentage of respondents (49.3%) reported using a non-privatized social networking website, compared to the 50.7% who either did not use a social networking website or privatized their website. An even larger percentage of college freshmen (58.5%) reported using a social networking website that was not privatized. Moreover, during both time periods, respondents reported posting various types of personal information on their social networking website and providing it to online contacts. These behaviors, which were examined further through bivariate and multivariate analysis, potentially could make respondents more suitable targets. Another notable behavior that could increase target suitability is communicating with strangers online. Nearly the same amount of college freshmen (n = 221) and high school seniors (n = 207) reported this communication, indicating that approximately half of the respondents in the study were willing to converse with people online whom they had never met previously.

There also were some notable reported frequencies regarding measures taken that indicated the presence or absence of capable guardianship. Perhaps not surprisingly, respondents indicated greater guardianship applied to their Internet use during the high school senior time period. The majority of respondents reported using the Internet at home as a high school senior (n = 447). Furthermore, they were more likely to have someone else in the room with them during Internet use (especially a parent or sibling), and also were more likely to have restrictions on time spent online and the viewing of adult websites.

In the same respect, high school seniors were more likely to report being actively monitored (n = 67) or having filtering and blocking software installed on their computers (n = 239). In contrast, a large number of college freshmen (n = 422) reported using the Internet mainly in their college dorm, likely a location with lessened supervision. To illustrate, a smaller number of college freshmen reported current active monitoring of Internet use (n = 13) or the known use of filtering and blocking software (n = 191). This is most likely a result of the independence acquired through moving out of the parent or guardian's home, which then lowers guardianship over Internet use.

Finally, with regard to online victimization, the study found that more respondents experienced self-reported victimization during the high school senior time period. Over 100 (22.8%) students as high school seniors reported experiencing the receipt of unwanted sexually explicit material, compared to only 48 (11%) during the college freshman time period. A larger amount of respondents reported experiencing the receipt of non-sexual harassment, with 144 (30.8%) receiving it as high school seniors and 52 (13%) during the college freshman time period. With regard to solicitation for sex, 45 (9.5%) reported receiving it as high school seniors, while 37 (7.75%) received it as college freshmen.

High school seniors were also more likely to participate in offline relationships with people met online. Eighty-five respondents during the high school senior time period (17.7%), versus 79 respondents (16.4%) during the college freshmen time period, participated in offline contact, which included activities such as speaking on the telephone or meeting the person at a home or different location. Moreover, 20 respondents (4.2%) during the high school senior time period and 12 (2.5%) respondents during the college freshmen time period reported participating in a willing sexual encounter with an online contact. Since college freshmen appear less supervised and reported greater Internet use, this finding may be surprising, as it might be assumed that college freshmen would be more likely to participate in offline relationships. However, perhaps because high school seniors are striving for independence and may be more naïve about participating in potentially dangerous behaviors, this finding might be explained as a consequence of rebellion or the assertion of independence while adolescents are living at home with their parents and family.

The majority of past studies in the area of online victimization of adolescents have presented frequency and descriptive data. The most recognized research on this topic comes from the Youth Internet Safety Survey (YISS), which is one of the most cited sources of data on Internet use and adolescent online victimization in the United States. With regard to online victimization, respondents in the current study experienced victimization at different rates than in either of two administrations of the YISS. Sixty-three percent of participants in the YISS-1 and 72% of participants in YISS-2, who were at least 15 years of age, reported experiencing unwanted exposure to sexually explicit material   (Mitchell et al., 2003). In the current study, only 22.8% of respondents as high school seniors and 10.0% as college freshman reported this type of online victimization. In contrast, when examining non-sexual online harassment, only 6% of the respondents in the YISS-1 and 9% of the respondents in the YISS-2 experienced this type of victimization, compared to over 30% of the respondents as high school seniors and 13% as college freshman in this study. Finally, with regard to sexual solicitation, respondents in both administrations of the YISS reported received this communication at a higher rate (13% and 19%, respectively) than in the current study (9.6% as high school seniors and 5.8% as college freshman).

It is quite possible that the age group of the respondents could account for the differences in victimization revealed. The YISS was based on a national sample of adolescents' ages 12 to 17 years old, while this study examined older adolescents (18 and 19 years old) from a single university in the northeast. Since more youth in the YISS reported experiencing victimization of a sexual manner (i.e., receipt of unwanted sexual material and sexual solicitation), it is possible that younger adolescents are more likely to engage in online behaviors (perhaps not measured in the current research) that increase their chances of this type of victimization. Alternatively, since the adolescents in this study reported higher rates of victimization in a non-sexual manner, it is possible that older adolescents are more likely to engage in online behaviors (again, perhaps not measured in the current research) that increase their likelihood of this victimization.

BIVARIATE CORRELATIONS

Bivariate correlations were also examined among the independent and dependent variables, separated again by time period in question (i.e., high school senior and college freshman) and each theoretical construct. First, the associations between the variables representing exposure to motivated offenders were examined. For both time periods, there were positive correlations between spending time on the Internet and use of CMCs. Therefore, high school seniors and college freshmen typically are likely to be using CMCs as part of their online activities. While high frequencies of respondents reported using MySpace and Facebook as social networking websites, a stronger correlation was found between the use of social networking websites and the use of

Facebook for college freshmen. This indicates that college freshmen are more likely to use Facebook over other types of social networking websites, which is not surprising since Facebook was originally created for college students.

There also were multiple significant correlations found between the variables representing target suitability. Communication with online contacts and the variables representing various types of information posted on a social networking website or given to an online contact (e.g., age, gender, and picture) were correlated for both time periods. Therefore, both high school seniors and college freshmen who communicate with people met online appear likely to provide personal information to these contacts or post it on their social networking websites. Moreover, respondents who reported that they provided personal information were likely to provide more than one type. In fact, 69% of high school seniors and 67% of college freshmen who revealed information on a social networking website posted at least 5 types of personal information. Furthermore, 59% of high school seniors and 61% of college freshmen who provided personal information to online contacts revealed at least 5 types of personal information. This supply of information would appear to increase target suitability, as subjects not only are conversing with strangers online, but also are revealing various pieces of personal information that could be used for targeted victimization.

There were only a few notable correlations between the guardianship variables. For both time periods, main use of a home computer was inversely associated with main computer use in a school computer lab. With the abundance of American households now owning a computer, this is not unexpected, as adolescents do not have to rely on school locations for computer use when they have the convenience of use at home. Other correlations reinforced the assumption that high school seniors would have more supervision during Internet use, as compared to college freshmen, since they typically are still living at home with parents or guardians. With regard to having the physical presence of someone in the room during Internet use, there was a positive correlation between main computer use at home and the presence of a parent or sibling in the room during the high school time period. As mentioned previously, high school seniors were more likely to use the Internet at home. College freshmen, on the other hand, were much more likely to live in a dorm, sharply decreasing the presence of supervisory adults.

The correlational examination of the dependent variables (i.e., types of victimization and formation of relationships) revealed similarities across time periods. High school seniors and college freshmen exhibited significant correlations between forming virtual relationships and participation in several types of offline contact, such as telephone calls and meeting an online contact in person. Essentially, no matter the time period, respondents who formed virtual relationships with people online were likely to carry those relationships offline. Participating in offline contact also was associated moderately with the experience of a willing sexual encounter.

Finally, there were some notable correlations between the independent/control variables and the dependent variables. These associations typically involved bivariate relationships between a dependent variable and the types of personal information provided to an online contact. This suggests that when adolescents increase their target suitability (through the revealing of personal information), they are more likely to be victimized online or form various types of relationships with people met online.

MULTIVARIATE FINDINGS AND HYPOTHESES

The purpose of this study was not simply to present frequencies and descriptive information about the sample in terms of Internet behaviors and experiences, but also to provide insight into what behaviors and activities influence online victimization and relationship formation. Hypothesis 1 focused on the effect of exposure to motivated offenders on the respondents' likelihood of victimization and relationship formation. Next, Hypotheses 2 through 4 considered how increasing a person's target suitability affects the likelihood of victimization and the formation of relationships with people met online. Finally, Hypotheses 5 through 7 allowed for an examination of the effect of protective measures (or lack thereof) on experiences with online victimization and relationship formation. A brief discussion of the explanatory power of the models is presented below, as well as a review of the results of the hypothesis tests and how the findings apply to Routine Activities Theory.

The explanatory power of the models was assessed through the computation and examination of the Cox & Snell and Nagelkerke R^2s. Overall, models estimated for the three types of victimization for the entire sample explained a relatively modest amount of variation in the dependent variables for both the high school senior and college freshman time periods. However, upon examination of the split models, the male models produced a respectable amount of explained variation, while the R^2s in the female models were rather small. With regard to the models estimated for the formation of various types of relationships with online contacts, a healthy amount of variation was explained for both time periods for the entire sample, as well as in the split models for males and females.

Further examination of the multivariate results indicated that certain behaviors that increased exposure to motivated offenders had a significant impact on the likelihood of victimization and relationship formation with online contacts. To illustrate, in support of Hypothesis 1 and consistent with the previous findings of Wolak et al. (2007), respondents in this study reported that participation in certain online activities and amplified used of CMCs increased the likelihood of victimization through receipt of unwanted sexually explicit material and non-sexual harassment. During the high school senior time period, for example, shopping online and using chat rooms increased the likelihood of receipt of unwanted sexually explicit material, while using the Internet to socialize and greater hours per week use of email was shown to increase the likelihood of receipt of non-sexual harassment. In particular, males who used chat rooms were significantly more likely to receive unwanted sexually explicit material, and males who used the Internet to socialize were more likely to receive non-sexual harassment.

Overall, supportive findings were observed for Hypothesis 1 for both the high school senior and college freshman time periods. These results, which indicate that exposure to motivated offenders increases a person's likelihood of experiencing victimization, also are consistent with those from more general victimization research testing Routine Activities Theory. Roncek and Maier (1991), for instance, found that a greater number of cocktail lounges and taverns on a residential block increased the likelihood of victimization in a particular area. Furthermore, research by Tewksbury and Mustaine (2000) concluded that greater exposure to potential offenders was associated with increased property crime victimization.

Multivariate analysis of the data also revealed that increases in target suitability had a large impact on the likelihood of victimization and forming relationships with online contacts. In fact, participating in behaviors that increased target suitability was shown to have the largest effect on dependent variables during both the high school senior and college freshman time periods. Supporting Hypothesis 2, as well as findings by Mitchell et al. (2007), results of this study indicated that communicating with people online and providing personal information to online contacts increased the likelihood of all three types of victimization during the high school senior time period. With regard to the college freshman time period, respondents who participated in these same behaviors also were more likely to receive non-sexual harassment.

In addition, respondents who provided personal information to online contacts were more likely to form various types of relationships with these online contacts, which supported Hypothesis 4. The results of the study also indicated that during both time periods, both male and female survey respondents who provided personal information to individuals met online were more likely to form virtual relationships, participate in some form of offline contact, and participate in sexual encounters.

These findings are comparable to those from previous research examining victimization and testing Routine Activities Theory. Multiple studies have found that lowered target suitability is associated with a decrease in the likelihood of becoming a victim of crime (Felson, 1996; Horney, et al., 1995; Schreck & Fisher, 2004). More specifically, Arnold et al. (2005) discovered that if the main activities of respondents involve drinking and other leisure activities, their level of target suitability is increased; in turn, they are more likely to be a victim of crime. Moreover, Wang's (2002) examination of factors associated with bank robberies revealed that banks presenting themselves as suitable targets (e.g., excessive amounts of cash and located close to a major highway) are more likely to be robbed.

Surprisingly, Hypothesis 3 was not supported in this study. Using a non-privatized social networking website, thought to increase a person's target suitability by exposing personal information to any viewer on the Internet, did not significantly influence the likelihood of victimization for high school seniors or college freshmen. The only dependent variable affected by this activity was forming a virtual relationship with a person met online. Both high school seniors and college freshmen who used a non-privatized social networking website were more likely to form virtual relationships.

Unlike the other two constructs of Routine Activities Theory, protective measures taken during Internet use (measured under the theoretical construct of lack of capable guardianship) had minimal effects on the dependent variables. With regard to the measures of lack of capable guardianship, findings from this study indicated that protective software had no significant effect on victimization for survey respondents. Contrary to what was predicted in Hypothesis 5, the use of filtering and blocking software did not appear to decrease victimization for the respondents. Conversely, respondents who were unsure if the software was present were more likely to be victimized. In other words, although there was a possibility that software was present to filter unwanted materials, respondents were still more likely to receive some type of unwanted sexual material.

Hypothesis 6 asserted that online restrictions given to respondents would decrease the likelihood of victimization. Limited support was found for this hypothesis, as only one type of restriction (viewing of adult websites) had a statistically significant effect on victimization. High school seniors who experienced this type of restriction were less likely to be victimized online, while college freshmen were not affected. Moreover, although not considered a form of victimization, restriction of the viewing of adult websites also decreased the likelihood of participating in face-to-face contact with a person met online.

Finally, mixed support was found for Hypothesis 7, which asserted that adolescents who were monitored while online were less likely to be victimized. Having a teacher or another person in the room did decrease the likelihood of receipt of non-sexual harassment for female respondents during both time periods. Furthermore, college freshmen who had a friend in the room were less likely to receive unwanted sexual material. However, male high school seniors with a parent in the room, as well as college freshmen with a person designated as "Other" in the room during Internet use, were more likely to receive unwanted sexual material or sexual solicitation.   As discussed previously, it is possible that an issue of temporal ordering was present, which influenced the direction of the findings. In other words, respondents could have first experienced the dependent variable, and as a result, later received greater monitoring in the room with them during Internet use.

In regard to the issue of temporal ordering, this explanation was offered at several points in this study to account for cause and effect relationships. For example, as cited earlier, logistic regression models indicated that the revocation of privileges by parents resulted in the increased likelihood of victimization for survey respondents. It is possible that the victimization occurred first, resulting in the revocation of privileges. However, there is a second possibility for the explanation of these relationships. Revocation of privileges could result in an adolescent turning to the Internet for entertainment, as all other privileges have been taken away (e.g., use of the car, socialization with friends, going out of the house). Rather than the previously offered explanation of temporal ordering, the actual victimization or formation of relationships with online contacts is a result of the increased use of the Internet.

As a further note, during both the high school senior and college freshmen time periods, a few of the dependent variables were influenced by the presence of a person designated as "Other." For example, female high school seniors were less likely to participate in offline contact with a person met online if they had a person designated as "Other" in the room with them during Internet use. Alternatively, college freshmen who had an "Other" person in the room were more likely to receive sexually explicit material and sexual solicitation.

On the survey, the options for persons present in the room during Internet use included parents, friends, siblings, and teachers; however, there was no opportunity on the questionnaire to provide a qualitative answer indicating who fell in the "Other" category. It could be assumed the "Other" category would include boyfriends/girlfriends and strangers or limited acquaintances, which would appear consistent with the findings. Strangers or limited acquaintances present in the room during the respondents' Internet use would offer little in the way of guardianship and could increase the likelihood of victimization for college freshman. On the other hand, the presence of a boyfriend in the room during Internet use could increase guardianship and potentially prevent a female high school senior from arranging offline contact with a person met online.

The findings of this study generally were different from results of previous studies examining victimization and testing Routine Activities Theory, as past research suggested that use of protective measures to increase guardianship was associated with a decreased likelihood of victimization (Cao & Maume, 1993; Cook, 1987; Sampson, 1987). In Wang's (2002) research mentioned previously, banks with greater security measures and armed security guards were less likely to be robbed. Tseloni et al. (2004) found further support using data from the British Crime Survey to examine burglary victimization. They discovered that single parent families were more likely to have their homes burglarized due to a lack of guardianship.

Overall, very few explanatory studies have been performed to examine the factors that affect the likelihood of online victimization for adolescents. However, in comparison to the limited research available on the topic, the findings of this study were somewhat similar to those of previous investigations. For example, Mitchell et al. (2007) found that adolescents who are female, use chat rooms, and talk to people online experience an increased likelihood of being solicited for sex while online. Similarly, in the current study, respondents reported at both the high school senior and college freshman time periods that communicating with people online and providing personal information to online contacts increased the likelihood of sexual solicitation. However, while use of chat rooms and gender influenced the likelihood of other types of victimization, they were not a significant predictor of sexual solicitation.

Additionally, Wolak et al. (2007) examined characteristics associated with exposure to sexual material online. Their analysis determined that various prevention characteristics (e.g., use of filtering software) , psychosocial characteristics (e.g., parent-child conflict), and Internet use characteristics were significantly associated with exposure to online pornography. Somewhat similar results were found in the current research. For instance, the use of chat rooms increased the likelihood of receipt of unwanted sexual material for high school seniors and college freshman. In addition, high school seniors who had privileges revoked by parents (i.e., parent-child conflict) were also more likely to be victimized. With regard to prevention characteristics, the presence of filtering software did not appear to have an affect on victimization, but having a friend in the room during Internet use significantly reduced the likelihood of victimization for college freshmen.

POLICY IMPLICATIONS

Past research has shown that adolescents are spending increasing amounts of time on the Internet (Addison, 2001; Izenberg & Lieberman, 1998; Nie & Erbring, 2000; United States Department of Commerce, 2002). This finding was confirmed in the current research, as respondents were found to have spent an average of 15.14 hours per week as high school seniors and 26.64 hours per week as college freshmen on the Internet, as well as sizeable amounts of time using email, instant messaging, chat rooms, and social networking websites. This augmented Internet use also comes with the risk of victimization, as confirmed in the current study and in past research. With this in mind, children, adolescents, and families need to be educated about the inherent risks associated with Internet use, in order to prevent victimization through safer online practices.

The findings of this study suggest that respondents who spend greater amounts of time using the Internet and specific CMCs (in turn increasing their likelihood of encountering a motivated offender) are more likely to be victimized. However, it would seem futile to attempt to develop stricter legislation or prevention programs that require or encourage youth to reduce their general use of the Internet. Use of the Internet is often necessary for educational purposes, and many youth successfully and safely use the Internet to socialize and connect with others. Based on the administration of the first Youth Internet Safety Survey, Wolak et al. (2002) reported that over half (55%) of the youth surveyed reported the use of chat rooms, instant messages, or email to communicate with people they had never met, with the hopes of forming relationships. Rather than trying to stop youth from socializing on the Internet, it would appear more effective to educate youth about the dangers present online and make them aware of the potential for victimization.

Children and adolescents using the Internet should be encouraged through educational programs to participate in online communication with people they know and trust, but be cautious about communicating and providing personal information to strangers. Many of the respondents in this study reported that they communicated with, and provided personal information to, people they met online, and some participated in offline encounters and relationships with these online contacts. In other words, these youth were revealing personal information to strangers (perhaps motivated offenders who may intend to prey upon a vulnerable population), and some were willing to extend virtual relationships offline through various modes of communication, including face-to-face contact. Although none of the respondents in this study reported participating in an unwilling sexual relationship with someone met online, past research has shown that there are adolescents who are physically victimized by contacts met on the Internet (Kendall, 1998; Tarbox, 2000). If youth limit online communication to people they know and only cautiously communicate with strangers, the risk of victimization should be lower.

Another popular suggestion for Internet victimization prevention is monitored Internet use. As reported by Volokh (1997), courts have suggested the use of filtering and blocking software as a protective measure for adolescents during Internet use. The Clinton administration endorsed use of this software, asserting it would protect youth better than any legislative statute (Clinton, 1997). Despite public support, studies testing the effectiveness of filtering and blocking software have found it to be an unreliable method of protection while online (Hunter, 2000; Mitchell et al., 2005). This was confirmed by the findings of the current study, as the use of filtering and blocking software was not shown to decrease the likelihood of online victimization.

Mitchell et al. (2005) hypothesized that parents may be more comfortable with a more active monitoring method rather than relying on technology. Although there was limited support for the hypotheses focused on lack of capable guardianship, a few specific guardianship variables were shown to be statistically significant predictors of the dependent variables. Unexpectedly, respondents who had a parent in the room during the college freshman time period were more likely to receive sexual solicitation from an online contact. However, this may an issue of temporal ordering, as respondents who experienced this type of outcome could be more likely to have a parent in the room monitoring Internet use as a result of the receipt of the sexual solicitation. Furthermore, parental monitoring might be more effective for children and younger adolescents, as compared to older youth. If parents are encouraged to monitor the time spent by their children on the Internet, websites they are visiting, and their child's online contacts, they could have an enhanced ability to prevent victimization beyond the protection provided by technology. This approach also may be more appealing to the children, if they are reassured that the parent/guardian is not monitoring Internet use due to lack of trust, but for the protection of the child.

Another alternative to parental monitoring that increases guardianship electronically is the use of Internet zoning. The Internet Community Ports Act (Preston, 2008), currently debated in legislative circles including the United States Congress, proposes the use of separated Internet ports to guard children against access to undesirable material. Based on the same premise as blocking children from adult-rated television programs, parents have the ability to give children access to various ports on the Internet, as well as to deny them access to ports that contain adult and vulgar material. "Community ports" would include online material that is not considered obscene or harmful, while "open ports" would include all types of material. By zoning the Internet into these specific types of material, parents can provide protection for their children against online predators, sexual material (including child pornography), and other unsavory forms of Internet communication.

No matter what the preferred solution by parents, the reality is that as children get older and become more independent adolescents, they are becoming more technologically savvy and therefore able to participate in online communication without the watchful eye of a parent or guardian. We as adults have the responsibility to educate youth that predators come in many forms, not just the stereotypical "creepy old man" preying on little children on the playground. This especially true on the Internet where multiple identities can be created and used to prey on young online users. The main goal we should have is to create not paranoia, but rather intelligent awareness.

Furthermore, it is also prudent that we not only focus on the victims of Internet crimes, but also the offenders. We are now debating what is the appropriate punishment for Internet crimes whose victims have not received tangible injuries from the offender. For example, what is the appropriate punishment for cyberstalking and how do you prove mental anguish that was supposedly caused from it? Is online sexual harassment really a crime when you can just erase the email? It is all these debatable issues that make current punishments for Internet crimes so hard to justify, and many are too lenient to deter offenders. If we want our children to stay safe, we have to fall back on the tenets of classical criminology and ensure effective punishment through the concepts of "swift, certain, and severe."

LIMITATIONS AND FUTURE RESEARCH

Little explanatory research has been conducted to examine the factors that influence the likelihood of online victimization for adolescents. Therefore, this study sought to make a significant contribution to a currently small body of literature. Although a variety of findings were uncovered, limitations to the current study should be recognized and future research performed in order to provide further understanding of the factors that contribute to the online victimization of youth. First, the limitations of this particular study will be examined, followed by a discussion of possibilities for future research.

Research Limitations

A sample of adolescents was chosen for this study because past research has shown that youth between the ages of 12 to 17 years old are at a high risk for online victimization (Mitchell, Finkelhor, & Wolak, 2003; O'Connell, Barrow, & Sange, 2002; Sanger, Long, Ritzman, Stofer, & Davis, 2004; Wolak, Finkelhor, & Mitchell, 2004; Wolak, Mitchell, & Finkelhor, 2002; Wolak, Mitchell, & Finkelhor, 2006). The ideal sample for this particular study would include respondents who fall into this age group. However, based on human subject issues that would have been encountered while trying to survey this group, college students who were legally able to participate in research (without parental consent) were chosen. The sample included adolescents ages 18 and 19 years old and was lacking the inclusion of younger adolescents. To address this situation, the sample of older adolescents were not only asked about current experiences, but also were asked to recall experiences from the high school senior time period.

Recalling accurate information from the past may be difficult for respondents, which in turn would affect the validity of the findings. Past studies, such as the YISS and research performed by Bremer and

Rausch (1998) and Rosen (2006), asked respondents questions regarding their current experiences and did not ask them to recall information from a previous time period. By asking questions that limit the scope of recall, the reliability and validity of the findings generally will be greater. On the other hand, this study asked about both current (college freshman time period) and past (high school senior time period) experiences and behaviors in an effort to address this limitation.

A second limitation regarding the representativeness of this sample is based on the geographical area from which the sample was drawn. The mid-sized university in the northeast is located in a rural area, and many of its students originate from surrounding rural areas. This limited the number of students from urban and suburban settings in the sample, thereby decreasing the generalizability of the findings. In comparison, the YISS-1 and YISS-2 utilized a nationally representative survey by collecting data from adolescents throughout the United States, making the results more generalizable. Nonetheless, since this is one of the few explanatory studies performed in this topic area, issues of recall and geographical location did not prevent a significant contribution from being made to the knowledge and understanding of potential causes of adolescent online victimization.

The final limitation of the research involves the way some of the questions were worded. First, while the survey provided comprehensive measurement of the three constructs of Routine Activities Theory and the dependent variables, there may have been issues associated with the temporal ordering of a few of the questions. For example, the findings indicated that respondents (during both time periods) who had privileges revoked by parents were more likely to be victimized. Also, college freshmen who had a parent in the room were more likely to receive sexual solicitation. As discussed previously, it is possible that close monitoring by parents, as well as revoked privileges, were a result of adolescent activities on the Internet. Wording of survey items to ensure proper temporal ordering, or the use of longitudinal research, would help address this limitation.

A second issue involving the wording of survey items pertains to the measurement of persons in the room with the respondent during Internet use. The variable representing having a person in the room designated as "Other" during Internet use was shown to be a significant predictor of multiple dependent variables, particularly for college freshmen and male respondents. However, a qualitative response to elaborate on the identity of the person designated as "Other" was not available in the survey. Since this was shown to be a significant independent variable, it should be investigated further in the future.

Suggestions for Future Research

As mentioned above, there may have been an issue associated with recall of accurate information regarding past experiences of respondents. A suggestion for future research would be to survey current high school, junior high school, and even elementary school students about their recent online experiences, to reduce the possible issue of accurate recall. The results of this research could be compared to other studies performed with adolescents, such as the Youth Internet Safety Survey, and would provide a more comprehensive assessment of the topic.

Future research should also survey respondents in different types of geographical settings. The mid-sized university in the northeast used in this study primarily contained a rural population; surveying students from urban and suburban settings could produce different results, as respondents from these areas may have different experiences with Internet use and victimization. Respondents also could be separated by age or grade level, to consider whether there are differences in predictors based on age.

Interestingly, use of a specific social networking website was not shown to significantly predict victimization among respondents in this study. However, use of MySpace and Facebook were both shown to increase the likelihood that a respondent (during the high school senior time period) would participate in a virtual relationship and face-to-face contact with a person met online. Since social networking websites are often used for the purpose of socialization and forming relationships, this is a logical finding. It would be beneficial to the literature to continue research in this specific area, to discover if these findings are externally valid. This research also could involve younger respondents and those in different geographical areas; similar findings would provide further support for cyber-awareness programs to educate youth about safety precautions to be used in the development of relationships with online contacts.

There was very little reported online offending revealed in this study, so no bivariate and multivariate analyses were conducted to examine potential relationships between independent variables and offending outcomes. It is possible the college students in this study actually did not engage in these behaviors, or there may have been under-reporting by the survey respondents. For future research, it may be beneficial to insert an additional clause reminding the respondents that all results are anonymous and will be kept confidential.

A final idea for future research is to include questions involving deception used online. Users of chat rooms and social networking websites can amend their true identity to serve their own purpose (Beebe et al., 2004; Freeman-Longo, 2000; Wolak et al., 2002), including to make themselves more appealing to younger Internet users. For example, a 40 year-old man with a fake identity could contact a 14 year-old girl via her MySpace page and attempt to participate in offline contact with her, which in turn could lead to the victimization of the female. Even if victimization is not involved, survey respondents who are consensual adults may have experiences with deceptive practices when meeting others online. If this is shown to be a substantial problem, preventative programs also could be developed to educate parents and adolescents about the dangers of deception online.

CONCLUSIONS

With limited past research available, this study sought to generate greater understanding about the relationships between Internet behaviors and activities (representing the three constructs of Routine Activities Theory) and online victimization and relationship formation. Providing personal information to online contacts and communicating with people met online (variables representing the theoretical construct of target suitability) were the strongest and most consistent predictors of online victimization, as well as the formation of relationships with people met online. Moreover, use of certain CMCs (variables representing the theoretical construct of exposure to motivated offenders) also was shown to predict certain types of victimization. However, variables representing the third construct of Routine Activities Theory, lack of capable guardianship, were not shown to be strong or consistent predictors of online victimization of youth.

From the knowledge gained through this study, hopefully more effective policies and programs can be developed to educate youth and families about protecting themselves while online. Youth should be aware of who they are conversing with online and refrain from providing any type of personal information to people they do not know and trust. Although this study did not indicate that the use of protective practices decreases the likelihood of victimization, more active monitoring of adolescents (particularly younger ones) might allow parents and guardians to be proactive in preventing victimization.

Finally, there is ample opportunity for future research in this area. Surveying a wider age range of adolescents, as well as those in different geographical areas, would add to the knowledge base. Also, further investigation of the use of social networking websites and the offending behaviors of adolescents, as well as their familiarity with deceptive Internet practices, will advance our knowledge of the online behaviors and experiences of adolescents. With this knowledge, better protective measures and policies can be developed to keep adolescents safe online.

Questionnaire

PART I

The questions below pertain to your own Internet use and experiences. Please answer all of the following questions regarding <u>your experiences as a high school senior</u>.

<u>AS A HIGH SCHOOL SENIOR:</u>

1. How many hours per day did you typically spend on the Internet?
_____

2. How many days per week did you typically use the Internet? _____

3. Please mark all the activities you performed on the Internet:

_____Research

_____Gaming

_____Planning travel

_____Website design

_____Shopping

_____Socializing with others

_____Other:_____

4. Did you use email as a high school senior? Please mark the appropriate answer.

_____Yes

_____No

4a. If you answered yes, how many hours per week did you typically spend using email? _____

5. Did you use instant messaging as a high school senior? Please mark the appropriate answer.

_____Yes

_____No

5a. If you answered yes, how many hours per week did you typically spend using instant messaging? _____

6. Did you use chat rooms as a high school senior? Please mark the appropriate answer.

_____Yes

_____No

6a. If you answered yes, how many hours per week did you typically spend using chat rooms? _____

7. Did you have a social networking website (i.e. MySpace/Facebook/Other) as a high school senior? Please mark the appropriate answer.

_____Yes

_____No

If you answered no to question 7, please skip to question 8.

7a. If you answered yes, how many hours per week did you typically spend using your social networking website? _____

7b. If you answered yes, where did you have a social networking website? Please mark all that apply:

_____MySpace

_____Facebook

_____Other:_____

7c. Was your social networking website marked "private," so only designated friends could see your profile? Please mark the appropriate answer.

_____Yes

_____No

7d. What types of information did you post on your social networking website? Please mark all that apply:

_____Age

_____Gender

_____Descriptive characteristics (e.g., hair color, eye color, weight)

_____Picture(s) of yourself

_____Telephone number

_____School location

_____Extracurricular activities

_____Goals/aspirations

_____Sexual information (e.g., fantasies, desires, experiences, etc.)

_____Emotional/mental distresses and problems (e.g., feelings of loneliness and sadness)

_____Family conflicts (e.g., problems with parents, arguments with siblings)

_____Other:_____

8. Did you communicate with people online, via email, instant messaging, or chat rooms, that you had never met in person? Please mark the appropriate answer.

_____Yes

_____No

9. Did you voluntarily give personal information to a person you met online? Please mark the appropriate answer.

_____Yes

_____No

9a. If you answered yes, please mark all the types of information that you gave to or discussed with a person you met online via **email, instant messaging, or chat rooms**:

_____Age

_____Gender

_____Descriptive characteristics (e.g., hair color, eye color, weight)

_____Picture(s) of yourself

_____Telephone number

_____School location

_____Extracurricular activities

_____Goals/aspirations

_____Sexual information (e.g., fantasies, desires, experiences, etc.)

_____Emotional/mental distresses and problems (e.g., feelings of loneliness and sadness)

_____Family conflicts (e.g., problems with parents, arguments with siblings)

_____Other:_____

10. Where did you most often use a computer? Please mark the **one location** used most often:

_____Parent/Guardian's Home

What room?

_____Living room/family room

_____Your bedroom

_____Parent/guardian's bedroom

_____Other_____

_____School computer lab

_____Friend's home

_____Coffee shop

_____Other:_____

11. Please mark any of the people listed who were typically in the same room when you used a computer. Mark all that apply.

_____Parent/Guardian

_____Friend

_____Teacher/Counselor

_____Sibling

_____Someone else

_____No One

12. Please mark all of the restrictions you had from your parent/guardian while using the Internet.

_____Time spent online

_____Viewing of adult websites (e.g., pornographic websites, gambling websites)

_____Use of email, instant messaging, chat rooms, and social
networking websites

_____Other:_____

_____No restrictions

13. To your knowledge, did your parent/guardian or computer software actively monitor your Internet use by regularly checking the websites you visited?

_____Yes

_____No

_____I don't know if someone was regularly monitoring my
Internet use

14. To your knowledge, was any type of blocking or filtering software on the computer(s) you typically used, to protect you from unwanted materials? Please mark the appropriate answer.

_____Yes

_____No

_____I don't know if blocking or filtering software was on my
computer

15. Did you receive unwanted sexually explicit material over the Internet (e.g., pornographic pictures of naked people or people having sex **NOT** including pop-ups)? Please mark the appropriate answer.

_____Yes

_____No

15a. If yes, on average, how many times per week did this occur?

_____

16. Did you send sexually explicit material to other people over the Internet (e.g., pornographic pictures of naked people or people having sex **NOT** including pop-ups)? Please mark the appropriate answer.

_____Yes

_____No

16a. If yes, on average, how many times per week did this occur?

_____

17. Were you harassed in a non-sexual manner on the Internet [e.g., repetitive, unwanted emails (including multiple spam mails from the same source) or instant messages]? Please mark the appropriate answer.

_____Yes

_____No

17a. If yes, on average, how many times per week did this occur?

_____

18. Did you harass others in a non-sexual manner on the Internet (e.g., repetitive, unwanted emails or instant messages)? Please mark the appropriate answer.

_____Yes

_____No

18a. If yes, on average, how many times per week did this occur?

_____

19. Were you asked for sex on the Internet? Please mark the appropriate answer.

_____Yes

_____No

19a. If yes, on average, how many times per week did this occur?

_____

20. Did you ask others for sex on the Internet? Please mark the appropriate answer.

_____Yes

_____No

20a. If yes, on average, how many times per week did this occur?

_____

21. Did you form any virtual relationships (i.e., a friendship or romantic relationship in which communication only occurred on the Internet) with people you met online? Please mark the appropriate answer.

_____Yes

_____No

22. Did you participate in any offline contact (i.e., communication not on the Internet) with someone you met on the Internet? Please mark the appropriate answer.

     _____Yes

     _____No

22a. If yes, please mark all forms of communication you participated in offline with a contact you met on the Internet.

     _____Mail via U.S. Postal Service

     _____Telephone conversation

     _____Received money or gifts

     _____Met at person's home

     _____Met at your home

     _____Met at other location

     _____Other:_____

22b. If yes to number 22 above, did the offline contact ever turn into a sexual encounter? Please mark the appropriate answer.

     _____Yes

     _____No

22c. If yes to 22b, did you participate willingly in sexual activity with the person? Please mark the appropriate answer.

     _____Yes

     _____No

PART II

The questions below pertain to your own Internet use and experiences. Please answer all of the following questions regarding <u>your experiences as a college freshman.</u>

<u>AS A COLLEGE FRESHMAN:</u>

1. How many hours per day did you typically spend on the Internet?

_____

2. How many days per week did you typically use the Internet? _____

3. Please mark all the activities you performed on the Internet:

     _____Research

_____Gaming

_____Planning travel

_____Website design

_____Shopping

_____Socializing with others

_____Other:_____

4. Did you use email as a college freshman? Please mark the appropriate answer.

_____Yes

_____No

4a. If you answered yes, how many hours per week did you typically spend using email? _____

5. Did you use instant messaging as a college freshman? Please mark the appropriate answer.

_____Yes

_____No

5a. If you answered yes, how many hours per week did you typically spend using instant messaging? _____

6. Did you use chat rooms as a college freshman? Please mark the appropriate answer.

_____Yes

_____No

6a. If you answered yes, how many hours per week did you typically spend using chat rooms? _____

7. Did you have a social networking website (i.e. MySpace/ Facebook/ Other) as a college freshman? Please mark the appropriate answer.

_____Yes

_____No

If you answered no to question 7, please skip to question 8.

7a. If you answered yes, how many hours per week did you typically spend using your social networking website? _____

7b. If you answered yes, where did you have a social networking website? Please mark all that apply:

_____MySpace

_____Facebook

_____Other:_____

7c. Was your social networking website marked "private," so only designated friends could see your profile? Please mark the appropriate answer.

_____Yes

_____No

7d. What types of information did you post on your social networking website? Please mark all that apply:

_____Age

_____Gender

_____Descriptive characteristics (e.g., hair color, eye color, weight)

_____Picture(s) of yourself

_____Telephone number

_____School location

_____Extracurricular activities

_____Goals/aspirations

_____Sexual information (e.g., fantasies, desires, experiences, etc.)

_____Emotional/mental distresses and problems (e.g., feelings of loneliness and sadness)

_____Family conflicts (e.g., problems with parents, arguments with siblings)

_____Other:_____

8. Did you communicate with people online, via email, instant messaging, or chat rooms, that you had never met in person? Please mark the appropriate answer.

_____Yes

_____No

9. Did you voluntarily give personal information to a person you met online? Please mark the appropriate answer.

_____Yes

_____No

9a. If you answered yes, please mark all the types of information that you gave to or discussed with a person you met online via **email, instant messaging, or chat rooms**:

_____Age

_____Gender

_____Descriptive characteristics (e.g., hair color, eye color, weight)

_____Picture(s) of yourself

_____Telephone number

_____School location

_____Extracurricular activities

_____Goals/aspirations

_____Sexual information (e.g., fantasies, desires, experiences, etc.)

_____Emotional/mental distresses and problems (e.g., feelings of loneliness and sadness)

_____Family conflicts (e.g., problems with parents, arguments with siblings)

_____Other:_____

10. Where did you most often use a computer? Please mark the **one location** used most often:

_____Parent/Guardian's Home

What room?

_____Living room/family room

_____Your bedroom

_____Parent/guardian's bedroom

_____Other: _____

_____School computer lab

_____Friend's home

_____Coffee shop

_____Other:_____

11. Please mark any of the people listed who were typically in the same room when you used a computer. Mark all that apply.

_____Parent/Guardian

_____Friend

_____Teacher/Counselor

_____Sibling

_____Someone else

_____No One

12. Please mark all of the restrictions you had from your parent/guardian while using the Internet.

_____Time spent online

_____Viewing of adult websites (e.g., pornographic websites, gambling websites)

_____Use of email, instant messaging, chat rooms, and social networking websites

_____Other:_____

_____No restrictions

13. To your knowledge, did your parent/guardian or computer software actively monitor your Internet use by regularly checking the websites you visited?

_____Yes

_____No

_____I don't know if someone was regularly monitoring my Internet use

14. To your knowledge, was any type of blocking or filtering software on the computer(s) you typically used, to protect you from unwanted materials? Please mark the appropriate answer.

_____Yes

_____No

_____I don't know if blocking or filtering software was on my computer

15. Did you receive unwanted sexually explicit material over the Internet (e.g., pornographic pictures of naked people or people having sex **NOT** including pop-ups)? Please mark the appropriate answer.

_____Yes

_____No

15a. If yes, on average, how many times per week did this occur?

_____

16. Did you send sexually explicit material to other people over the Internet (e.g., pornographic pictures of naked people or people having sex **NOT** including pop-ups)? Please mark the appropriate answer.

_____Yes

_____No

16a. If yes, on average, how many times per week did this occur?

_____

17. Were you harassed in a non-sexual manner on the Internet [e.g., repetitive, unwanted emails (including multiple spam mails from the same source) or instant messages]? Please mark the appropriate answer.

_____Yes

_____No

17a. If yes, on average, how many times per week did this occur?

_____

18. Did you harass others in a non-sexual manner on the Internet (e.g., repetitive, unwanted emails or instant messages)? Please mark the appropriate answer.

_____Yes

_____No

18a. If yes, on average, how many times per week did this occur?

_____

19. Were you asked for sex on the Internet? Please mark the appropriate answer.

_____Yes

_____No

19a. If yes, on average, how many times per week did this occur?

_____

20. Did you ask others for sex on the Internet? Please mark the appropriate answer.

_____Yes

_____No

20a. If yes, on average, how many times per week did this occur?

_____

21. Did you form any virtual relationships (i.e., a friendship or romantic relationship in which communication only occurred on the Internet) with people you met online? Please mark the appropriate answer.

_____Yes

_____No

22. Did you participate in any offline contact (i.e., communication not on the Internet) with someone you met on the Internet? Please mark the appropriate answer.

_____Yes

_____No

22a. If yes, please mark all forms of communication you participated in offline with a contact you met on the Internet.

_____Mail via U.S. Postal Service

_____Telephone conversation

_____Received money or gifts

_____Met at person's home

_____Met at your home

_____Met at other location

_____Other:_____

22b. If yes to number 22 above, did the offline contact ever turn into a sexual encounter? Please mark the appropriate answer.

_____Yes

_____No

22c. If yes to 22b, did you participate willingly in sexual activity with the person? Please mark the appropriate answer.

_____Yes

_____No

PART III

Please answer the following questions about your own personal characteristics:

1. Class Standing (please mark the appropriate answer):

_____Freshman (0–28 credits)

_____Sophomore (29–56 credits)

_____Junior (57-90 credits)

_____Senior (91 credits or more)

2. Sex (please mark the appropriate answer):

_____Male

_____Female

3. Age:_____

4. Please mark the choice that best portrays your race/ethnicity:

_____White Non-Hispanic

_____White Hispanic

_____American Indian or Alaska Native

_____African-American

_____Asian

_____Other

5. Please mark your current living situation:

_____Living with parent(s)/guardian(s)

_____Living with other family member(s) besides parent(s)

_____Living in a dormitory

_____Living in a rented apartment/house

_____Living in a Greek fraternity/sorority house

_____Other:_____

6. Please mark your general academic performance at the **time of high school graduation**:

_____All A's

_____Mostly A's and B's

_____Mostly B's

_____Mostly B's and C's

_____Mostly C's

_____Mostly C's and D's
_____Mostly D's
_____Mostly D's and F's

Regarding your personal feelings <u>when you were a high school senior</u>, please indicate how much you disagree or agree with the following statements by placing a vertical slash mark on the line below each statement. PLEASE DO NOT SIMPLY CIRCLE ONE OF THE CHOICES AT THE TWO ENDS OF THE LINE. Rather, draw a vertical line on the continuum where it most accurately represents your disagreement or agreement with each statement.

Here is a hypothetical example, for which you would probably place your slash mark toward the right side of the line.

A. I like going to the beach for summer vacation.

Strongly Disagree   Strongly Agree

As a high school senior:

1. I could share my thoughts and feelings with my parents/guardians.

Strongly Disagree   Strongly Agree

2. I could share my thoughts and feelings with my friends.

Strongly Disagree   Strongly Agree

3. I enjoyed spending time with my friends.

Strongly Disagree   Strongly Agree

4. I had respect for my parents.

Strongly Disagree   Strongly Agree

5. I had respect for my teachers.

Strongly Disagree   Strongly Agree

6. Participation in school activities was important to me.

_____

Strongly Disagree   Strongly Agree

7. I tried to stay involved in activities at school.

_____

Strongly Disagree   Strongly Agree

8. Getting good grades was important to me.

_____

Strongly Disagree   Strongly Agree

9. I tried hard to succeed at school.

_____

Strongly Disagree   Strongly Agree

10. My parents/guardians often yelled at me.

_____

Strongly Disagree   Strongly Agree

11. My parents/guardians often nagged me.

_____

Strongly Disagree   Strongly Agree

12. My parents/guardians often took away privileges.

_____

Strongly Disagree   Strongly Agree

Regarding your personal feelings <u>currently as a college student at IUP</u>, please indicate how much you disagree or agree with the following statements by placing a vertical slash mark on the line below each statement. PLEASE DO NOT SIMPLY CIRCLE ONE OF THE CHOICES AT THE TWO ENDS OF THE LINE. Rather, draw a vertical line on the continuum where it most accurately represents your disagreement or agreement with each statement.

Here is a hypothetical example, for which you would probably place your slash mark toward the right side of the line.

A. I like going to the beach for summer vacation.

_____

Strongly Disagree   Strongly Agree

As a college freshman:

1. I can share my thoughts and feelings with my parents/guardians.

_____

Strongly Disagree   Strongly Agree

2. I can share my thoughts and feelings with my friends.

_____

Strongly Disagree   Strongly Agree

3. I enjoy spending time with my friends.

_____

Strongly Disagree   Strongly Agree

4. I have respect for my parents.

_____

Strongly Disagree   Strongly Agree

5. I have respect for my teachers.

_____

Strongly Disagree   Strongly Agree

6. Participation in school activities is important to me.

_____

Strongly Disagree   Strongly Agree

7. I try to stay involved in activities at school.

_____

Strongly Disagree   Strongly Agree

8. Getting good grades is important to me.

_____

Strongly Disagree   Strongly Agree

9. I try hard to succeed at school.

_____

Strongly Disagree   Strongly Agree

10. My parents/guardians often yell at me.

_____

Strongly Disagree   Strongly Agree

11. My parents/guardians often nag me.

_____

Strongly Disagree   Strongly Agree

12. My parents/guardians often take away privileges.

_____

Strongly Disagree   Strongly Agree

THIS IS THE END OF THE SURVEY.
THANK YOU FOR YOUR PARTICIPATION IN THIS STUDY.

Informed Consent Form

You are invited to participate in this research study. The following information is provided in order to help you to make an informed decision whether or not to participate. If you have any questions, please do not hesitate to ask. You are eligible to participate because you are a student at Indiana University of Pennsylvania (IUP); however, you must be <u>at least 18 years old</u> to participate.

The purpose of this study is to examine the Internet use of IUP students, as well as their experiences with online victimization. You will be asked to complete a survey asking you questions about the following: frequency of Internet use, different methods of communication you use on the Internet, information provided to others on the Internet, and experiences regarding unwanted non-sexual harassment, exposure to sexual material, and sexual solicitation online. Also, you will be asked about your experiences with forming relationships with online contacts and if those relationships formed into any offline relationships. Participation in this study will require approximately 30 minutes of your time and is not considered a part of your coursework. Participation or non-participation will not effect the evaluation of your performance in this class.

The information gained from this study will help the researcher better understand the use of the Internet by college students, as well as the frequency with which they experience victimization through various methods. All answers will be kept completely anonymous. No identifying information (i.e., name, birth date, student ID #) will be requested in the survey, so all information you provide will be anonymous.

Your participation in this study is voluntary. You are free to decide not to participate in this study or to withdraw at any time without adversely affecting your relationship with the investigators or IUP. Your decision will not result in any loss of benefits to which you are otherwise entitled. If you choose to participate, you may withdraw at any time by writing the word withdraw on the front of your survey. The researcher will destroy your survey. If you choose to participate, all information will be held in strict confidence and will have no bearing on your academic standing or services you receive from the University. Your responses will be considered only in combination with those from other participants. The information obtained in the study may be published in scientific journals or presented at scientific meetings, but your identity will be kept strictly confidential.

If you suffer emotional distress from participation in this study, a sheet containing contact information for various mental health providers is attached.

If you are willing to participate in this study, please tear this Informed Consent Form off the attached survey and keep for your own files. If you have any questions regarding the survey, please feel free to contact the researchers below:

This project has been approved by the Indiana University of Pennsylvania Institutional Review Board for the Protection of Human Subjects (Phone: 724/357-7730).

Counseling Services

Center for Counseling and Psychological Services
Indiana University of Pennsylvania
Pratt Hall, Room 307
201 Pratt Drive
Indiana, PA 15705
Telephone: 724-357-2621

Indiana County Guidance Center
793 Old Route 119 Highway North
Indiana, PA 15701
Telephone: 725-465-5576

Bibliography

Addison, D. (2001). Youngsters increase time online as sites seek return visits. *Marketing, 10.*

Agresti, A. & Finlay, B. (1997). *Statistical methods for the social sciences* (3rd ed.). Upper Saddle River, NJ: Prentice-Hall.

Armagh, D. (1998). A safety net for the Internet: Protecting our children. *Juvenile Justice, 5(1),* 9-15.

Armstrong, G., & Griffing, M. (2007). The effect of local life circumstances on victimization of drug-involved women. *Justice Quarterly, 24(1),* 80-105.

Arnett, J. (1992). Reckless behavior in adolescence: A developmental perspective. *Developmental Review, 12,* 339-373.

Arnold, R., Keane, C., & Baron, S. (2005). Assessing risk of victimization through epidemiological concepts: An alternative analytic strategy applied to Routine Activities Theory. *Canadian Review of Sociology & Anthropology, 42(3),* 345-364.

Ashcroft v. American Civil Liberties Union, 535 U.S. 564 (2002), 217 F.3d 162.

Bachmann, R., & Schutt, R. (2007). *The practice of research in criminology and criminal justice* (10th ed). Thousand Oaks, CA: Sage Publications.

Bandura, A. (1997). *Self-efficacy: The exercise of control.* New York: Freeman.

Bandura, A. (2000). Social cognitive theory: An agentic perspective. *Annual Review of Psychology, 52,* 1-26.

Beebe, T., Asche, S., Harrison, P, & Quinlan, K. (2004). Heightened vulnerability and increased risk-taking among adolescent chat room users: Results from a statewide school survey. *Journal of Adolescent Health,35(2)*, 116-123.

Bendel, R. & Afifi, A. (1977). Comparison of stopping rules in forward regression. *Journal of the American Statistical Association, 72,* 46-53.

Bennett, R. (1991). Routine activities: A cross-national assessment of a criminological perspective. *Social Forces, 70,* 143-163.

Benthin, A., Slovis, P., & Severson, H. (1993). A psychometric study of adolescent risk perception. *Journal of Adolescence, 16,* 153-168.

Bernburg, J., & Thorlindsson, T. (2001). Routine activities in social context: A closer look at the role of opportunity in deviant behavior. *Justice Quarterly, 18(3),* 543-567.

Brame, R., Paternoster, R., Mazerolle, P., & Piquero, A. (1998). Testing for the equality of maximum-likelihood regression coefficients between two independent equations. *Journal of Quantitative Criminology, 14(3),* 245-261.

Brantingham, P., & Brantingham, P. (1981). *Environmental criminology.* Beverly Hills, CA: Sage.

Bremer, J., & Rauch, P. (1998). Children and computers: Risks and benefits. *Journal of the American Academy of Child and Adolescent Psychiatry, 37(5),* 559-561.

Cao, L., & Maume, D. (1993). Urbanization, inequality, lifestyles and robbery: A comprehensive model. *Sociological Focus, 26(1),* 11-26.

Carmines, E., & Zeller, R. (1979). *Reliability and validity assessment.* Thousand Oaks, CA: Sage Publications.

Catsambis, S. (1994). The path to math: Gender and racial-ethnic differences in mathematics participation from middle school to high school. *Sociology of Education, 67,* 199-215.

Children's Internet Protection Act, Public Law 106-554, 2000.

Clemmitt, M. (2006). Cyber socializing. *CQ Researcher, 16(27),* 1-34.

Clinton, W. (1997, July 16). *Remarks by the president at event on the e-chip for the Internet.* Retrieved from The White House Office of the Press Secretary [Online], September 1, 2005, from: http://www.whitehouse.gov/WH/News/Ratings/ remarks.html

Cohen, J. (1988). *Statistical power analysis for the behavioral sciences.* Hillsdale, NJ: Lawrence Erlbaum Associates, Publishers.

Cohen, L., & Cantor, D. (1980). The determinants of larceny: An empirical and theoretical study. *Journal of Research in Crime and Delinquency, 17(2),* 140-159.

Cohen, L., & Felson, M. (1979). Social change and crime rate trends: A routine activity approach. *American Sociological Review, 44,* 588-608.

Cohen, L., & Felson, M. (1981). Modeling crime trends: A criminal opportunity perspective. *Journal of Research in Crime and Delinquency, 18,* 138-164.

Collins, J., Cox, B., & Langan, P. (1987). Job activities and personal crime victimization: Implications for theory. *Social Science Research, 16,* 345-360.

Cook, P. (1987). Robbery violence. *Journal of Criminal Law and Criminology, 78,* 357-376.

Cooper, A. (1997). The Internet and sexuality: Into the next millennium. *Journal of Sex Education and Therapy, 22(1),* 5-6.

Danet, B. (1998). Revisiting computer-mediated communication and community. *Cybersociety 2.0,* 129-158.

Dean, S. (2006). *Sexual predators: How to recognize them on the Internet and on the street; how to keep your kids away.* Los Angeles: Silver Lake Publishing.

Deleting Online Predators Act, HR 5319 IH (2006).

DeVellis, R. (2003). *Scale development: Theory and applications (2nd ed.).* Thousand Oaks, CA: Sage Publications.

Dillman, D. (2007). *Mail and Internet surveys: The tailored-design method 2007 update with Internet, visual, and mixed-mode guide (2nd ed.).* Hoboken, NJ: John Wiley & Sons, Inc.

Durkin, K. (1997). Misuse of the Internet by pedophiles: Implications for law enforcement and probation practice. *Federal Probation, 61(3)*, 14-19.

Durkin, K., & Bryant, C. (1999). Propagandizing pederasty: A thematic analysis of the on-line exculpatory accounts of unrepentant pedophiles. *Deviant Behavior: An Interdisciplinary Journal, 20*, 103-127.

Durkin, K., Wolfe, T., & Clark, G. (1999). Social bond theory and binge drinking among college students: A multivariate analysis. *College Student Journal, 33*, 450-461.

Ehrhardt-Mustaine, E., & Tewksbury, R. (1997). The risk of victimization in the workplace for men and women: An analysis using Routine Activities/Lifestyle Theory. *Humanity & Society, 21(1)*, 17-38.

Elmer-DeWitt, P., & Bloch, H. (1995). On a screen near you: Cyberporn. *Time, 146(1)*, 38-45.

Fact Sheet: The Adam Walsh Protection and Safety Act of 2006. (2006, July 27). Retrieved October 20, 2006, from WhiteHouse.gov: http://www.whitehouse.gov/news/releases/2006/07/20060727-7.html

Federal Communications Commission (2006). *Children's Internet Protection Act.* Retrieved December 30, 2006, from http://www.fcc.gov/cgb/consumerfacts/ ipa.html

Felson, M. (1986). Linking criminal choices, routine activities, informal social control, and criminal outcomes. In D. Cornish and R. Clarke (Eds.), *The Reasoning Criminal* (pp. 119-128). New York: Springer-Verlag.

Felson, M. (1987). Routine activities and crime prevention in the developing metropolis. *Criminology, 25*, 911-932.

Felson, M. (1994). Crime and everyday life: Insight and implications for society. Thousand Oaks, CA: Pine Forge Press.

Finn, J. (2004). A survey of online harassment at a university campus. *Journal of Interpersonal Violence, 19(4)*, 468-483.

Fitzpatrick, M. (2006, June 10). Testimony before house energy and commerce subcommittee on oversight and investigations.

Flanagin, A. (2005). IM online: Instant messaging use among college students. Communication Research Reports, 22(3), *175-187.*

Fleming, M., S. Greentree., D. Cocotti-Muller, K. Elias, and S. Morrison. (2006). "Safety in Cyberspace: Adolescents' Safety and Exposure Online." *Youth and Society* 38: 135-154.

Fowler, Jr., F. (2002). *Survey research methods.* Thousand Oaks, CA: Sage Publications.

Forde, D., & Kennedy, L. (1997). Risky lifestyles, routine activities,and the general theory of crime. *Justice Quarterly, 14(2),* 265-289.

Freeman-Longo, R. (2000). Children, teens and sex on the Internet. *Sexual Addiction & Compulsivity*, 7, 75-90.

Frisbie, D., Fishbine, G., Hintz, R., Joelsons, M., & Nutter, J. (1977). *Crime in Minneapolis: Proposals for prevention.* St. Paul, Minn.: Governor's Commission on Crime Prevention and Control.

Gaetz, S. (2004). Safe streets for whom? Homeless youth, social exclusion, and criminal victimization. *Canadian Journal of Criminology and Criminal Justice, 46(4),* 423-455.

Gallo, D. (1998). Filtering tools, education, and the parent: Ingredients for surfing safely on the information superhighway. *The APSAC Advisor, 11(4)*, 23-25.

Garofalo, J. & Clark, D. (1992). Guardianship and residential burglary. Justice Quarterly, 9(3), 443-463.

Garofalo, J., Siegel, L., & Laub, J. (1987). School-related victimizations among adolescents: An analysis of National Crime Survey narratives. Journal of Quantitative Criminology, 3, 321-338.

Gibson, W. (1984). *Neuromancer.* New York: Ace Books.

Giedd, J., Blumenthal, J., Jeffries, N., Castellanos, F., Liu, H., Zijdenbos, A., Paus, T., Evans, A., & Rapoport, J. (1999). Brain development during childhood and adolescence: A longitudinal MRI study. *Nature Neuroscience, 2(10),* 861-863.

Goodson, P. McCormick, D. & Evans, A. (2000). Sex and the Internet: A survey instrument to assess college students' behavior and attitudes. *CyberPsychology & Behavior, 3(2)*, 129-149.

Gore, A. (1999, May 5). *Remarks on the Internet.* Retrieved from The White House Office of the Press Secretary [Online], September 1, 2005, from: http://www.whitehouse.gov/WH/News/html/19990505-4219.html

Graham, J. (2003, October 20). Instant messaging program are no longer just for messages. *USA Today*, pp. 5D

Hair, J.F., Anderson, R.E., Tatham, R.L., & Black, W.C. (1998). *Multivariate data analysis* (5th ed.). Upper Saddle River, NJ: Prentice Hall.

Hardy, M., & Bryman, A. (2004). *Handbook of Data Analysis.* Thousand Oaks, CA: Sage Publications.

Harris Interactive. (2001). *Presenting: The class of 2001.* Retrieved September 2, 2005, from: http://www.harrisinteractive.com/news/allnewsbydate.asp?NewsID=292

Hawley, A. (1950). *Human ecology.* New York: The Ronald Press Company.

Hawdon, J. (1996). Deviant lifestyles: The social control of routine activities. *Youth and Society, 28,* 162-188.

Henderson, H. (2005). *Internet predators.* New York: Facts On File.

Hindelang, M., Gottfredson, M., & Garofalo, J. (1978). *Victims of personal crime: An empirical foundation for a theory of personal victimization.* Cambridge, MA: Ballinger Publishing Company.

Hirschi T. (1969). *Causes of delinquency.* Berkeley and Los Angeles: University of California Press.

Horney, J., Osgood, D., & Marshall, I. (1995). Criminal careers in the short-term: Intra-individual variability in crime and its relation to local life circumstances. *American Sociological Review, 60,* 655-673.

Hunter, C. (2000). Social impacts: Internet filter effectiveness testing – over- and underinclusive blocking decisions of four popular web filters. *Social Science Computer Review, 18(2), 214-223.*

Huttenlocher, P. (1979). Synaptic density in human frontal cortex developmental changes and effects of aging. *Brain Research, 2(16),* 195-205.

Izenberg, N., & Lierbman, D. (1998). The web, communication trends, and children's health: How the children use the web. *Clinical Pediatrics, 37(6),* 335-340.

Jackson, A., Gililand, K., & Veneziano, L. (2006). Routine activity theory and sexual deviance among male college students. *Journal of Family Violence, 21(7),* 449-460.Jones, S. (1999). *Doing Internet research.* London: Sage.

Kandell, J. (1998). Internet addiction on campus: The vulnerability of college students. *CyberPsychology and Behavior, 1(1),* 11-17.

Kendall, V. (1998). The lost child: Congress's inability to protect our teenagers. *Northwestern University Law Review, 92(4),* 1307-1315.

Kennard, W. (1999, May 4). *Remarks of William Kennard at the Annenberg Public Policy Center conference on Internet and the family [Online].* Retrieved September 1, 2005, from: http://www.fcc.gov/Speeches/Kennard/spwek916.html

Kennedy, L. & Forde, D. (1990). Risky lifestyles and dangerous results: Routine activities and exposure to crime. *Sociology & Social Research, 74(4),* 208-211.

Kirkpatrick, M. (2006, May 17). Top 10 social networking sites see 47 percent growth. *Thesocialsoftwareweblog.* Retrieved September 25, 2006, from http://socialsoftware.weblogsinc.com

Kleinrock, L. (1961). Information flow in large communication net. *RLE Quarterly Progress Report.*

Kleinrock, K. (1964). *Communication nets: Stochastic message flow and delay.* New York: McGraw-Hill.

Kuder, G., & Richardson, M. (1937). The theory of the estimation of test reliability. *Psychometrika, 2,* 151-160.

LaGrange, T. (1999). The impact of neighborhoods, schools, and malls on the spatial distribution of property damage. *Journal of Research in Crime and Delinquency, 36(4),* 393-422.

Lamb, M. (1998). Cybersex: Research notes on the characteristics of the visitors to online chat rooms. *Deviant Behavior, 19(2),* 121-135.

Lamb, A. & Johnson, L. (2006). Want to be my friend? What you need to know about social technologies. *Teacher Librarian, 34(1), 55-57.*

Lasley, J. (1989). Drinking routines/lifestyles and predatory victimization: A causal analysis. *Justice Quarterly, 6(4),* 529-542.

Lea, M., & Spears, R. (1995). Love at first byte? Building personal relationships over computer networks. In S. Duck (Ed.), *Understudied relationships: Off the beaten track* (pp. 197-233). Thousand Oaks, CA: Sage.

Lebo, H. (2000). *The UCLA Internet report: Surveying the digital future.* Los Angeles: UCLA Center for Communication.

Leiner, B., Cerf, V., Clark, D., Kahn, R., Kleinrock, L., Lynch, D., Postel, J., Roberts, L. & Wolff, S. (2003). A brief history of the Internet. *Internet Society.* Retrieved September 24, 2006, from http://www.isoc.org/internet/history/brief.shtml

Lenhart, A., Rainie, L., & Lewis, O. (2001). *Teenage life online: The rise of the instant-message generation and the internet's impact on friendships and family relationships.* Washington, DC: Pew Internet & American Life Project. Retrieved September 25, 2006, from www.pewinternet.org/reports/toc.asp?Report=36

Leonard, K., & Decker, S. (1994). The theory of social control: Does it apply to the very young? *Journal of Criminal Justice, 22(2),* 89-105.

Lewis-Beck, M. (1980). *Applied regression: An introduction.* Thousand Oaks, CA:Sage Publications.

Lorig, K., Laurent, D., Deyo, R., et al. (2002). Can a back pain email discussion group improve health status and lower health care costs? A randomized study. *Archives of Internal Medicine, 162,* 792-796.

Lwin, M., A. Stanaland, and A. Miyazaki. 2008. "Protecting Children's Privacy Online: How Parental Mediation Strategies Affect Website Safeguard Effectiveness." *Journal of Retailing* 84: 205-217.

Lynch, J. (1987). Routine activity and victimization at work. *Journal of Quantitative Criminology, 3,* 283-300.

Madriz, E. (1996). The perception of risk in the workplace: A test of routine activity theory. *Journal of Criminal Justice, 24(5),* 407-412.

Marriott, M. (1998, July 2). The blossoming of Internet chat. *The New York Times,* pp. G1.

McAuliffe, W. (2001). *Court drops hammer on "Wonderland" child-porn participants.* Retrieved January 31, 2003, from ZDNet: http://zdnet.com/2100-11-528167.html?legacy=zdnn

McCabe, K. (2000). Child pornography and the Internet. *Social Science Computer Review, 18(1),* 73-76.

McFarlane, M., Bull, S.S. & Rietmeijer, C. (2000). The Internet as a newly emerging risk environment for sexually transmitted diseases. *Journal of the American Medical Association, 284,* 443-446.

Medaris, M. & Girouard, C. (2002). *Protecting children in cyberspace: The ICAC task force program.* Washington, DC: National Center for Missing & Exploited Children.

Meier, R. & Miethe, T. (1993). Understanding theories of criminal victimization. In M. Tonry (Ed.), *Crime and Justice: An Annual Review of Research* (pp. 459-499). Chicago: University of Chicago Press.

Menard, S. (2007). *Applied logistic regression analysis* (2nd ed.). Thousand Oaks, CA:Sage Publications.

Messner, S. & Blau, J. (1987). Routine leisure activities and rates of crime: A macro-level analysis. *Social Forces, 65,* 1035-1052.

Messner, S. & Tardiff, K. (1985). The social ecology of urban homicide: An application of the routine activities approach. *Criminology, 23(2),* 241-267.

Meyers, L., Gamst, G., & Guarino, A. (2005). *Applied multivariate research: Design and interpretation.* Thousand Oaks, CA: Sage Publications.

Miethe, T., & Meier, R. (1990). Opportunity, choice, and criminal victimization: A test of a theoretical model. *Journal of Research in Crime and Delinquency, 27(3),* 243-266.

Miethe, T. & McDowall, D. (1993). Contextual effects in models of criminal victimization. Social Forces, 71, 741-759.

Miethe, T., Stafford, M., & Long, J. (1987). Social differentiation in criminal victimization: A test of routine activities/lifestyles theories. *American Sociological Review, 52,* 184-194.

Miller v. California, 413 U.S. 15; 93 S. Ct. 2607; 37 L. Ed. 2d 419

Mitchell, K., Finkelhor, D., & Becker-Blease, K. (2007). Linking youth Internet and conventional problems: Findings from a clinical perspective. *Journal of Aggression, Maltreatment, & Trauma, 15(2),* 39-58.

Mitchell, K., Finkelhor, D. & Wolak, J. (2003). The exposure of youth to unwanted sexual material on the Internet: A national survey of risk, impact and prevention. *Youth & Society, 34(3),* 3300-3358.

Mitchell, K., Finkelhor, D., & Wolak, J. (2005). Protecting youth online: Family use of filtering and blocking software. *Child Abuse & Neglect,* 29(7), 753-765.

Mitchell, K., Finkelhor, D., & Wolak, J. (2007). Youth Internet users at risk for the more serious online sexual solicitations. *American Journal of Preventative Medicine, 32(6),* 532-537.

Mitchell, K., Wolak, J., & Finkelhor, D. (2005). Police posting as juveniles to catch sex offenders: Is it working? *Sexual Abuse: A Journal of Research and Treatment, 17,* 241-267.

Moriarty, L., & Williams, J. (1996). Examining the relationship between routine activities theory and social disorganization: An analysis of property crime victimization. *American Journal of Criminal Justice, 21(1),* 43-59.

Mota, S. (2002). The U.S. Supreme Court addresses the Child Pornography Prevention Act and Child Online Protection Act in

Ashcroft v. Free Speech Coalition and *Ashcroft v. American Civil Liberties Union. Federal Communications Law Journal, 55,* 85-98.

Mustaine, E., & Tewksbury, R. (1999). A routine activities theory explanation for women's stalking victimization. *Violence Against Women, 5(1),* 43-62.

Mustaine, E., & Tewksbury, R. (2002). Sexual assault of college women: A feminist interpretation of a routine activities analysis. *Criminal Justice Review, 27(1),* 89-123.

MySpace.com. (2006). *MySpace.com.* Retrieved November 1, 2006, from http://www.myspace.com

Nie, N., & Erbring, L. (2000). *Internet and society: A preliminary report.* Stanford: Stanford Institute for the Quantitative Study of Society.

O'Connell, R., Barrow, C., & Sange, S. (2002). *Young peoples use of chat rooms: Implications for policy strategies and programs of education.* Retrieved November 1, 2005, from http://www.uclan.ac.uk/host/cru/publications.htm

Olivia, R. (2003). Instant messaging comes of age. *Marketing Management, 12(3),* 49.

Osborn, C. (2006, June 20). Teen, mom sue MySpace.com for $30 million. *Austin-American Statesman.*

Osgood, D., Wilson, J., O'Malley, P., Bachman, J., & Johnston, J. (1996). Routine activities and individual deviant behavior. *American Sociological Review, 61,* 636-55.

Parry, W. (2006, October 19). Feds net 125 nationwide in kid-porn case. *Yahoo!News.* Retrieved October 19, 2006, from: http://news.yahoo.com/s/ap/20061019/ ap_on_re_us/child_porn_ arrests&printer=1

Paternoster, R., Brame, R., Mazerolle, P., & Piquero, A. (1998). Using the correct statistical test for the equality of regression coefficients. *Criminology, 36(4),* 859-866.

Pew Research. (2001). *Teenage life online: The rise of the instant message generation and the Internet's impact on friendships and family relationships 2001*. Retrieved September 1, 2006, from: http://www.pewinternet.org/reports/pdfs/ PIP_Teens_Report.pdf

Pew Research. (2003). *America's online pursuits: The changing picture of who's online and what they do.* Retrieved September 1, 2006, from: http://www.pewinternet.org/ pdfs/PIP_Online_Pursuits_ Final.pdf

Preston, C. (2008). Zoning the Internet: A new approach to protecting children online. *Brigham Young University Law Review*, 1417-1469.

PRNewswire.com. (2006, June 15). *Social networking sites continue to attract record numbers as MySpace.com surpasses 50 million U.S. visitors in May.* RetrievedJune 15, 2006, from: http://www.PRNewswire.com

Punch, K. (2003). *Survey research: The basics.* Thousand Oaks, CA: Sage Publications.

Quayle, E., & Taylor, M. (2003). Model of problematic internet use in people with a sexual interest in children. *CyberPsychology & Behavior, 6, 93-106.*

Radicati Group. (2003). *Instant messaging and presence market trends, 2003-2007.* Retrieved October 1, 2006, from http://www.research andmarkets.com/eportinfo.asp?report_id=35248&t=e&cat_id=4

Rainie, L. (2006). *Life online: Teens and technology and the world to come.* Speech to the annual conference of the Public Library Association, Boston. Retrieved October 1, 2006, from www.pewinternet.org/ppt/ Teens%20and%20technology.pdf

Ramirez, A., Dimmick, J., & Lin, S. (2004). *Revisiting media competition: The gratification niches of instant messaging, email, and telephone.* Paper presented at the Annual Meetings of the International Communication Association, New Orleans.

Reeves, P. (2000). Coping in cyberspace: The impact of Internet use on the ability of HIV-positive individuals to deal with their illness. *Journal of Health Communication, 6,* 47-59.

Rhodes, W., & Conly, C. (1981). Crime and mobility: An empirical study. In P. Brantingham & P. Brantingham, *Environmental Criminology* (pp. 167-188). Beverly Hills, CA: Sage.

Roberts, L. (1967, October). *Multiple computer networks and intercomputer communication.* Paper presented at the meeting of the Association of Computing Machinery Conference, Gatlinburg, TN.

Roberts, D., Foehr, U., Rideout, V., & Brodie, M. (1999). Kids & media @ the new millennium: A comprehensive analysis of children's media use. *The Henry J. Kaiser Family Foundation.*

Roncek, D., & Bell, R. (1981). Bars, blocks, and crimes. *Journal of Environmental Systems, 11,* 35-47.

Roncek, D., & Maier, P. (1991). Bars, blocks, and crimes revisited: Linking the theory of routine activities to the empiricism of "hot spots." *Criminology, 29(4),* 725-753.

Rosen, L. (2006). *Adolescents in MySpace: Identity formation, friendship and sexual predators.* California State University, Dominguez Hills.

Rosenbaum, M., Altman, D., Brodie, M. Flournoy, R., Blendon,R. & Benson, J. (2000). NPR/Kaiser/Kennedy School Kids & Technology Survey. *Retrieved September 24, 2006 from http://www.npr.org/programs/specials/pool/technology /technology.kids.html*

Sampson, R. (1987). Personal violence by strangers: An extension and test of the opportunity model of predatory victimization. *Journal of Criminal Law and Criminology, 78,* 327-356.

Sampson, R., & Wooldredge, J. (1987). Linking the micro- and macro-dimension of lifestyle-routine activity and opportunity models of predatory victimization. *Journal of Quantitative Criminology, 3,* 371-393.

Sanger, D., Long, A., Ritzman, M., Stofer, K. & Davis, C. (2004).Opinions of female juvenile delinquents about their interactions in chat rooms. *Journal of Correctional Education, 55(2),* 120-131.

Sasse, S. (2005). "Motivation" and routine activities theory. *Deviant Behavior, 26,* 547-570.

Schnarch, D. (1997). Sex, intimacy and the Internet. *Journal of Sex Education and Therapy, 22(1),* 15-20.

Schreck, C. & Fisher, B. (2004). Specifying the influence of the family and peers on violent and victimization. *Journal of Interpersonal Violence, 19(9),* 1021-1041.

Schwartz, M., DeKeseredy, W., Tait, D., & Alvi, S. (2001). Male peer support and a feminist routine activities theory: Understanding sexual assault on the college campus. *Justice Quarterly, 18(3),* 623-649.

Schwartz, M. & Pitts, V. (1995). Exploring a feminist routine activities approach to explaining sexual assault. *Justice Quarterly, 12,* 9-31.

Sherman, L., Gartin, P., & Buerger, M. (1989). Hot spots of predatory crime: Routine activities and the criminology of place. *Criminology, 27(1),* 27-55.

Simon, B. (1999). United States v. Hilton. *Berkeley Technology Law Journal, 14(1),* 385-401.

Simon, J. (2006). Computer-mediated communication: Task performance and satisfaction. *Journal of Social Psychology, 146(3),* 349-379.

Sloan, III, J.J. (1994). The correlates of campus crime: An analysis of reported crimes on college and university campuses. *Journal of Criminal Justice, 22(1),* 51-61.

Spano, R. & Nagy, S. (2005). Social guardianship and social isolation:

An application and extension of Lifestyle/Routine Activities Theory to rural adolescents. *Rural Sociology, 70(3),* 414-437.

Steinberg, L. (2004). Risking-taking in adolescence: What changes, and why? *Annals of the New York Academy of Sciences, 1021,* 51-58.

Stevens, J. (1992). *Applied multivariate statistics for the social sciences* (2nd ed.). Hillsdale, NJ: Lawrence Erlbaum Associates.

Stewart, E. (2003). School social bonds, school climate, and school misbehavior: A multilevel analysis. *Justice Quarterly, 20(3),* 575-604.Stutzman, F. (2006). *Social networking on campus.* University of North Carolina, Chapel Hill.

Tarbox, K. (2000). *Katie.com.* New York: Penguin Group. The Free Speech Coalition v. Reno, No. C 97-0281 VSC, 1997 WL 487758, at *7 (N.D. Cal. Aug. 12, 1997)

The guide. (2001). *On Magazine, 6(6).* Retrieved October 1, 2006, from: http://www.GetNetWise.com

The Henry J. Kaiser Family Foundation. (2000). *National survey of American adults on technology and national security: American kids on technology.* Menlo, CA: National Public Radio, Kaiser Family Foundation, et al.

Tewksbury, R. & Mustaine, E. (2000). Routine activities and vandalism: A theoretical and empirical study. *Journal of Crime & Justice, 23(1),* 81-110.

Tita, G. & Griffiths, E. (2005). Traveling to violence: The case for a mobility-based spatial typology of homicide. *The Journal of Research in Crime and Delinquency, 42(3),* 275-308.

Tseloni, A., Wittebrood, K., Farrell, G., & Pease, K. (2004). Burglary victimization in England and Wales, the United States and the Netherlands. *The British Journal of Criminology, 44(1),* 66-91.

Turi, L. (1997). Scholarly communication through electronic mailing. *Monist, 80(3),* 472-478.

Turkle, S. (1995). *Life on the screen: Identity in the age of the Internet.* New York: Simon & Schuster.

United States Department of Commerce. (2002). Computer and Internet usage by age and disability status: 2002. Washington, DC: United States Department of Commerce. Retrieved March 1, 2007, from http://www.census.gov/hhes/www/ disability/sipp/disab02/ ds02f6.pdf

United States Department of Justice. (2006, May 17). *Project Safe Childhood.* Retrieved November 3, 2006, from http://www.projectsafechildhood.gov/

United States v. Bodenheimer, D.C. No. 03-596 MCA (D.N.M. 2005), *aff'd* 05-2221 (10th Cir. 2006).

United States v. Hilton, 999 F. Supp. 131, 134 (D. Me. 1998)

United States v. X-Citement Video, 513 U.S. 64 (1994)

Virginia Tech. (1997, July 21). *"Three-prong obscenity test."* Retrieved November 12, 2006, from the Virginia Tech website: http://courses.cs.vt.edu/~cs3604/lib Censorship/3-prong-test.html

Vogel, R. & Himelein, M. (1995). Dating and sexual victimization: An analysis of risk factors among precollege women. *Journal of Criminal Justice, 23,* 153-162.

Volokh, E. (1997). Freedom of speech, shielding children, and transcending balancing. *Supreme Court Review, 141,* 141-197 [Online]. Retrieved September 1, 2005, from: http://www.law.ucla.edu/faculty/volokh/shield.htm

Wang, J. (2002). Bank robberies by an Asian gang: An assessment of the Routine Activities Theory. *International Journal of Offender therapy and Comparative Criminology, 46(5),* 555-568.

Wang, Q. & Ross, M. (2002). Differences between chat room and email sampling approaches in Chinese men who have sex with men. *AIDS Education Prevention, 14,* 361-366.

Whelan, D. (2001). The instant messaging market. *American Demographics, 23(12),* 28-31.

Wigfield, A., Eccles, J., Schiefele, U., Roeser, R., & Davis-Kean, P. (2006). Development of achievement of motivation. In W. Damon & N. Eisenberg (Eds.), *Handbook of child psychology* (pp. 933-1002), New York: Wiley.

Wolak, J., Mitchell, K., Finkelhor, D. (2002). Close online relationships in a national sample of adolescents. *Adolescence, 37(147),* 441-455.

Wolak, J., Mitchell, K.J., & Finkelhor, D. (2003). Escaping or connecting? Characteristics of youth who form close online relationships. *Journal of Adolescent Health, 26,* 105-119.

Wolak, J., Mitchell, K.J., & Finkelhor, D. (2004). Internet-initiated sex crimes against minors: Implications for prevention based on findings from a national study. *Journal of Adolescent Health, 35(5),* 11-20.

Wolak, J., Mitchell, K.J., & Finkelhor, D. (2006). *Online victimization of children: Five years later.* Washington, DC: National Center for Missing & Exploited Children.

Wolak, J., Mitchell, K.J., & Finkelhor, D. (2007). Unwanted and wanted exposure to online pornography in a national sample of youth Internet users. *Pediatrics, 119(2),* 247-257.

Wooldredge, J., Cullen, F., & Latessa, E. (1992). Victimization in the workplace: A test of routine activities theory. *Justice Quarterly, 9,* 325-335.

Ybarra, M., Mitchell, K., Finkelhor, D., & Wolak, J. (2007). Internet prevention messages: Targeting the right online behaviors. *Archives of Pediatric and Adolescent Medicine, 161,* 138-145.

Index